Engineering Geology

工程地质学

Feng Jinyan　Chen Jun　Yao Yangping
冯锦艳　陈　军　姚仰平

北京航空航天大学出版社

Abstract

The basic principle and application of Engineering Geology are expounded systematically in this book, which is divided into six chapters, namely, geological process and geological time, rock forming minerals and rocks, rock stratum and structures, weathering, river and groundwater, common geological disasters and engineering geological investigation. Contents are systematic and easy to be mastered. Thinking questions are provided in each chapter.

This book is written for undergraduate and postgraduate students of civil engineering. It is also a good reference book for qualifying examinations, such as Civil Engineering Qualifying Examination, Geotechnical Engineering Qualifying Examination and Environmental Assessment Qualifying Examination.

图书在版编目(CIP)数据

工程地质学 = Engineering Geology:英文 / 冯锦艳,陈军,姚仰平编著. -- 北京 : 北京航空航天大学出版社,2016.11

ISBN 978-7-5124-2307-7

Ⅰ. ①工… Ⅱ. ①冯… ②陈… ③姚… Ⅲ. ①工程地质一英文 Ⅳ. ①P642

中国版本图书馆 CIP 数据核字(2016)第 266808 号

Engineering Geology

工程地质学

Feng Jinyan　Chen Jun　Yao Yangping

冯锦艳　陈　军　姚仰平

责任编辑　冯　颖

*

北京航空航天大学出版社出版发行

北京市海淀区学院路 37 号(邮编 100191)　http://www.buaapress.com.cn

发行部电话:(010)82317024　传真:(010)82328026

读者信箱:emsbook@buaacm.com.cn　邮购电话:(010)82316936

北京建宏印刷有限公司印装　各地书店经销

*

开本:710×1 000　1/16　印张:8.75　字数:186 千字

2016 年 11 月第 1 版　2024 年 8 月第 3 次印刷

ISBN 978-7-5124-2307-7　定价:29.00 元

Preface

As noted in the preface of *Engineering Geology* written by F. G. Bell, engineering geology can be defined as the application of Geology to engineering practice. In other words, it is concerned with those geological factors that influence the location, design, construction and maintenance of engineering works. Accordingly, it draws on a number of geological disciplines such as geomorphology, structural geology, sedimentology, petrology and stratigraphy. In addition, engineering geology involves hydrogeology and some understanding of rock and soil mechanics. This book is written for undergraduate and postgraduate students of civil engineering. It is hoped that this book will also be of value to those involved in the profession.

It is a regret that this book can not cover all the needs of the variety of readers who may use it. Therefore, references are provided for those who want to pursue some aspect of the subject matter to greater depth. Obviously, students of civil engineering will have done much more reading on engineering geology than the basic geological material. Moreover, this book will reflect the background of its authors and their views of the subject. However, its authors have attempted to give a balanced overview of the subject. The first, second and third chapters are compiled by Feng Jinyan, the fourth chapter is compiled by Yao Yangping, and the fifth and sixth chapters are compiled by Chen Jun.

The authors of this book gratefully acknowledge all those who have given permission to publish material from other sources. Meanwhile, the authors would like to acknowledge the National Natural Science Funds of China (Grant No. 41302273) and the support of the School of Transportation Science and Engineering, Beihang University and all of the readers.

Beihang University
Feng Jinyan
April, 2016

Contents

Introduction

Geology is an important component of Earth Science, which has developed into a branch and consists of two categories of theoretical system by the 1980s. One category is to discuss the basic subjects (such as Petrology, Mineral Deposit Geology, Dynamic Geology, Structural Geology, Geomorphology and so on), and the other category is to research the interdisciplines (such as Geophysics, Geology Mechanics, Hydrogeology, Geochronology, Engineering Geology, Disaster Geology and so on). Engineering Geology, which is a branch of Geology, is to research geological problems related to human's engineering and construction activities.

China is a country with an ancient civilization, which has built many large projects as early as the Spring and Autumn Period. All of these suggest that the ancient Chinese people not only have superb construction techniques, but also know the engineering geological environment of building sites.

After the First World War, the whole world began to enter a large-scale construction period. In 1929, Karl Terzaghi, who is an Austrian, published the first book of *Engineering Geology* in the world. In the 1930s, the former Soviet Union geologists put forward a complete and systematic engineering geology and a theory system. In 1935, Savarenski Fiodor Petrovich (1881—1946) established the engineering geological and hydrogeological research laboratory in Institute of Geological Exploration in Moscow (Московский геологоразведочный институт), and published *Hydrology Geology* (1933) and *Engineering Geology* (1937), which marked the birth of Engineering Geology. Savarenski Fiodor Petrovich put forward and developed the natural history viewpoint of Engineering Geology. R. F. Legget wrote the book of *Geology and Engineering* in 1939 and published another great work of *Civil Engineering Geology Manual*. Austrians named J. Stini and L. Müller J. are the earliest people to realize the influence of structural plane on the rock mass , and established the journal of *Geology and Civil Engineering* in 1951.

Engineering Geology in China began to develope only after the founding of the People's Republic of China. In the early 1950s, Ministry of Geology established Bureau of Geology and the corresponding research institutions with the needs of national defense and economic construction, and set the Hydrological Geology to cultivate professional personnel.

Many major projects in that time promoted the rapid development of Engineering Geology in China, such as Sanmenxia Reservoir, Wuhan Yangtze River Bridge, and Xin'anjiang Hydroelectric Power Station. At the same time, some new engineering geological theories and ideas had been formed. Mr Gu Dezhen proposed structure control theory in the research of the stability of rock mass, and published a book of *Basement of Rock Engineering Geomechanics*. Mr Liu Guochang pointed out the research direction of regional stability on the basis of regional engineering geological condition and published a book of *China Regional Engineering Geology*. Mr Hu Haitao inherited and developed geomechanics theory from Mr Li Siguang, and published an article named *Regional Stability Analysis and Evaluation on Guangdong Nuclear Power Station Site Selection and Planning* combined with the thought to find a relatively stable "safety island" in an active region for the site selection of large-scale projects. In recent years, Engineering Geology in China developes very rapidly and keeps synchronizing with the world, which has formed its own feature. For example, in 2014, Mr Huang Runqiu completed the project named as "Geological Hazard Assessment and Prevention of Wenchuan Earthquake" and put forward the distribution regularity of seismic geological disaster dominated by "the seismogenic fault effect" and "topography effect".

Human activity is closely related to the engineering geological environment. Bad behavior will cause massive destruction of geological environment, for example, impoundment of reservoirs may induce earthquake, and over-exploitation of groundwater will cause ground subsidence of the city. The influence of human activities on the geological environment has reached or exceeded the degree of some certain natural geological function. Mr Wang Sijing considers that Environmental Geology is a new branch of Engineering Geology in terms of its research objects and theoretical basis, whose innovation point is to emphasize the effect and role of human engineering activities on the environment. Along with the increasing scale of environmental engineering geological problems and the enlarging influence range, China issued Registration of EIA Engineer, whose object is to assess and evaluate the influence of all the construction and planning on environment.

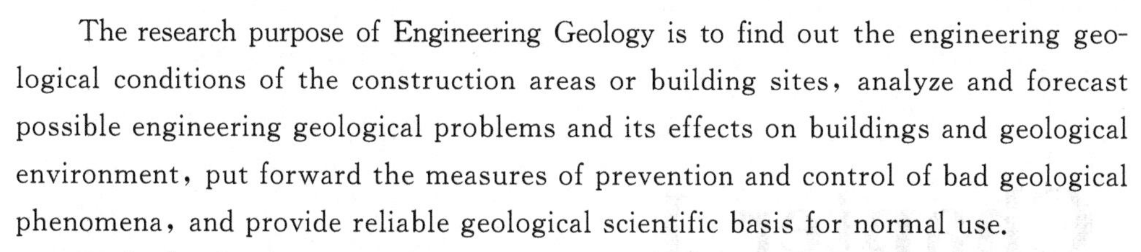

The research purpose of Engineering Geology is to find out the engineering geological conditions of the construction areas or building sites, analyze and forecast possible engineering geological problems and its effects on buildings and geological environment, put forward the measures of prevention and control of bad geological phenomena, and provide reliable geological scientific basis for normal use.

With the development of economic construction, massive infrastructure projects will be built, such as high-grade highway, harbor wharf, bridge, tunnel, airports subway, reconstruction project and many high-rise buildings, which will bring many new research topics. Geological workers need to develop new theories, new methods, and new technologies to overcome problems, and promote the development of engineering geology subject further.

This book is designed for civil engineering and airport road engineering, which have a wide range of contents, focus of the outstanding and the rich practice. It forms a system with Soil Mechanics, Rock Mechanics and Foundation Engineering, Construction Technology and other related courses. This book is divided into six chapters, namely, geological process and geological time, rock forming minerals and rocks, rock stratum and structures, weathering, river and groundwater, common geological disasters and engineering geological investigation.

Through the study of this course, engineering geology exploration task, content and method will be understood, engineering geological problems will be analyzed, the influence of human's engineering activities on the geological environment will be evaluated , and the corresponding countermeasures and management measures will be put forward.

Chapter 1

Geological Process and Geological Time

1.1 The Earth

The Earth is made up of different state material layers, namely crust, mantle and core (shown in Figure 1.1).

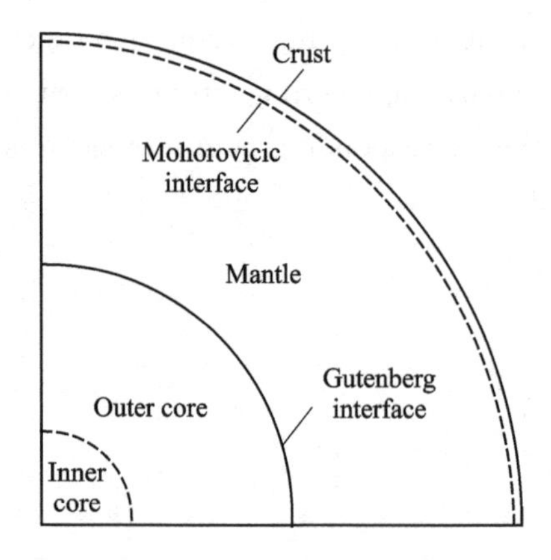

Figure 1.1 Interior structure of the Earth

1.1.1 Crust

The Earth's crust is a very thin and uneven layer, whose average thickness is about 17 km. The thickness of continental crust can amount to 70 km and the average thickness is about 33 km. Qinghai-Tibet Plateau is the thickest place of the

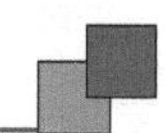

crust, whose thickness is more than 70 km. Oceanic crust is relatively thin, whose thickness is 5 ~ 10 km and the average thickness is about 7 km.

The Earth's crust consists of two layers described in below:

- The upper layer is sialic layer, also known as granite layer.
- The lower layer is sima, also called basaltic layer.
- Sialic layer and sima are separated by Conrad discontinuity.

The average density of the Earth's crust is about 2.8 g/cm^3, the sialic layer is about 2.7 g/cm^3 and the sima is about 2.9 g/cm^3. Sialic layer at the bottom of the ocean is very thin, especially ocean pelvic floor area. Sialic layer is missing in the middle of the Pacific, which is a discontinuity layer. The sima distributes on the continents and oceans, which is a continuous layer.

1.1.2 Mantle

The interface between the Earth's crust and the mantle is termed as the Mohorovicic (Moho) surface, which is found by seismologist Andrija Mohorovicic of Yugoslavia in 1909. He found that some seismic waves arrived at observation station faster than expected, the speed of longitudinal wave increased from 7.0 km/s to 8.1 km/s suddenly, and the shear wave velocity increased from 4.2 km/s to 4.4 km/s under the Earth's crust. This is because there is an interface between the crust and the mantle, namlhy Moho surface.

The mantle's range is from the Mohorovicic surface to the depth of 2 900 km, which accounts for about 83.3% of the Earth's volume. The mantle is divided into two layers according to the change of seismic wave. The depth of upper mantle is from 33 km to 980 km, which is composed by ultrabasic rocks. The depth of lower mantle is from 980 km to 2 900 km, which is composed by silicate, metallic oxide and sulfide. The density of lower mantle is about 5.1 g/cm^3.

Lithosphere includes the Earth's crust and the top of the upper mantle, which is on the asthenosphere. The asthenosphere is located in the upper portion of the upper mantle—of which the temperature is up to about 1 300 ℃ and the pressure is about 30 000 atmospheres. It is called asthenosphere because it can move slowly with half a viscous state. The asthenosphere is likely to be the birthplace of magma.

In 1914, German geophysicist Gutenberg (He became an American in 1936) found that longitudinal wave velocity decreased and shear wave disappeared at the underground of about 2 900 km depth. Later, it turned out that there is an interface between the core and the mantle, which is known as the Gutenberg interface.

1.1.3 Centrosphere

The centrosphere is the core part of the Earth and below the Earth's mantle. The centrosphere includes three layers, namely the outer core, the transition layer and the inner core.

The range of the outer core is from 2 900 km to 4 700 km below the Earth's surface, which is mainly composed by the mixture of molten iron and nickel, and a small amount of light elements such as Si and S. The density of the outer core is about 10 g/cm^3. The range of inner core is from 5 100 km to 6 371 km below the Earth surface, which is mainly composed by heavy metals such as iron, nickel.

The density of the inner core is about 12.5 g/cm^3.

The transition layer is located between the outer core and the inner core, whose thickness is 400 km. The state of matter of the transition layer is from liquid to solid.

1.1.4 Information Statistics of the Earth Structure

Table 1.1 is the information statistics of the Earth structure.

Table 1.1 Information statistics of the Earth structure

<table>
<tr><th colspan="4">Name of the Earth's spheres</th><th rowspan="2">Depth/km</th><th rowspan="2">Seismic P-wave velocity/(km · s^{-1})</th><th rowspan="2">Seismic shear wave velocity/(km · s^{-1})</th><th rowspan="2">Density/(g · cm^{-3})</th><th rowspan="2">State of matter</th></tr>
<tr><th>The first classification</th><th colspan="2">The second classification</th><th>Traditional classification</th></tr>
<tr><td rowspan="3">Ectosphere</td><td colspan="2">Crust</td><td>Crust</td><td>0~33</td><td>5.6~7.0</td><td>3.4~4.2</td><td>2.6~2.9</td><td>Solid</td></tr>
<tr><td rowspan="2">Outside tran-sition layer</td><td>Upper</td><td>Upper mantle</td><td>33~980</td><td>8.1~10.1</td><td>4.4~5.4</td><td>3.2~3.6</td><td>Part of the molten material</td></tr>
<tr><td>Bottom</td><td>Lower mantle</td><td>980~2 900</td><td>12.8~13.5</td><td>6.9~7.2</td><td>5.1~5.6</td><td>Liquid—Solid</td></tr>
<tr><td>Liquid layer</td><td colspan="2">Liquid layer</td><td>Outer Core</td><td>2 900~4 700</td><td>8.0~8.2</td><td>It cannot pass</td><td>10.0~11.4</td><td>Liquid</td></tr>
<tr><td rowspan="2">Endosphere</td><td colspan="2">Inside transition layer</td><td>Transition layer</td><td>4 700~5 100</td><td>9.5~10.3</td><td></td><td>12.3</td><td>Liquid—Solid</td></tr>
<tr><td colspan="2">Centrosphere</td><td>Inland nuclear</td><td>5 100~6 371</td><td>10.9~11.2</td><td></td><td>12.5</td><td>Solid</td></tr>
</table>

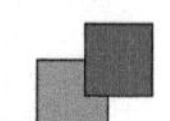

1.2 Geological Process

Geological process refers to the changes of the Earth's crust material composition, internal structure and surface morphology caused by the natural power. Some geological processes occur very quickly such as earthquake and vulcanian eruption, and some geological processes are very slow, for example, the Dutch coast goes down 2 mm per year on average.

The power of geological process comes from the internal thermal energy produced by metamorphosis of radioactive elements of the Earth's interior, solar radiation, the rotating force of the Earth and gravity. Geological process is divided into endogenic geological process and exogenic geological process according to the position and the origin of the energy.

1.2.1 Endogenic Geological Process

Endogenic geological process refers to the changes of material composition, structure, form of the lithosphere caused by the Earth's interior energy, which are divided into crustal movement, magmatism, earthquake action and metamorphism.

1. Crustal Movement

Crustal movement (tectonic movement) includes lifting and sunken of the Earth's crust or lithosphere, the outline changes of the land and the sea, the formation and development of mountain trench, etc. According to the motion directions, crustal movement can be divided into horizontal motion and lifting movement.

Horizontal motion refers that the Earth's crust or lithosphere block moves along the horizontal direction, such as adjacent blocks' separation, meeting or shearing, which is the most intense crustal evolution process. It is generally believed that horizontal motion is the main reason for the formation of various tectonic on the Earth's crust surface, such as fold, fracture or basin. San Andreas Fault in the United States and the Himalayas in China are the products of the horizontal motions.

Lifting movement refers that the Earth's crust moves perpendicular to the surface, namely along the Earth's radius direction, which causes regression and transgression of the sea and large uplift and settlement.

Table 1.2 provides the comparison between horizontal motion and lifting movement.

Table 1.2　Horizontal motion vs lifting movement

Crustal movement	Motor direction	Rock performance	Motion result
Horizontal motion	Material of the Earth's crust moves along the horizontal direction	Rocks bend and uplift, or break and open	Huge folded mountain, rift valley and ocean
Lifting movement	Material of the Earth's crust moves perpendicularly to the surface of the Earth	The Earth's crust rises or drops	The Earth's surfac is up and down, and the exchange appears between the land and the sea

2. Magmatism

Magmatism is that the active magma intrudes or even erupts out of the Earth surface along weak zones due to great pressure, which is divided into vulcanian eruption and intrusion (refer to Subsection 2.4).

3. Metamorphism

Metamorphism refers that the matter, structure or texture of the rocks has changed under the conditions of high temperature, high pressure and participation of other chemicals (refer to Subsection 2.3).

4. Earthquake Action

Earthquake action refers to the rapid vibration phenomenon of the Earth's crust, most of which are caused by crustal movement and named tectonic earthquake (refer to Subsection 5.4).

1.2.2　Exogenic Geological Process

Exogenic geological process refers to the geological process caused by the energy beyond the Earth. The energy mainly comes from solar radiation and the gravitational pull of the Sun and the Moon.

Exogenic geological process is divided into weathering, denudation, transportation, deposition and diagenesis.

Weathering refers to physical and chemical change processes on site of the Earth's crust under the conditions of the ambient temperature and pressure.

Denudation refers to the peeling of the weathering products from rocks, destruction of unweathered rocks and change of the appearance of rocks. It is divided

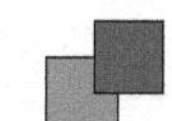

into wind erosion, running water erosion, groundwater submarine erosion, dig erosion of glacier and water erosion of lake.

Transportation makes the weathered and denuded products leave parent rocks under the action of force and then arrive in the sedimentary area through a long distance. The force mainly comes from wind, surface water, glaciers, underground water, lake water and ocean water.

Deposition makes the transported material separate from the transportation medium and form sediment because of the weaken momentum of the transportation medium, the changes of the physical and chemical conditions of the transportation medium and the biological role after a certain distance.

Diagenesis refers to the process that loose sediment becomes sedimentary rocks, which includes pressure effect, cementation and recrystallization.

1.3 Geologic Time

A particular rock unit required a certain interval of time to be formed. Hence, stratigraphy not only deals with strata but also with age, and the relationship between strata and age.

Accordingly, time units and time-rock units have been recognized. Time units are simply intervals of time, the largest ones of which are eons, although this term tends to be used infrequently. There are two eons, representing Pre-Cambrian Time and Phanerozoic Time.

Eons are divided into eras, and eras into periods (shown in Table 1.3). Periods are, in turn, divided into epochs and epochs into stages.

Time units and time-rock units are directly comparable, that is, there is a corresponding time-rock unit for each time unit.

The time-rock unit corresponding to an eon is an eonothem, corresponding to an era is a erathem, corresponding to a period is a system, corresponding to an epoch is a series, corresponding to a stage is a stage, and corresponding to achron is a chronozone.

Indeed, the time allotted to a time unit is determined from the rocks of the corresponding time-rock unit.

Table 1.3 Geologic Time Scale

Relative age						Actual age/ (million years)	Start time of biology		Major characteristic
Eon	Era		Period		Epoch		Plant	Animal	
Phanerozoic Time	Cenozoic (Kz)		Quaternary		Holocene(Q_4)	0.01	Modern plants	Human beings appeared	Many kinds of modern debris and glaciers appeared and loess generated
					Pleistocene(Q_{1-3})	2.5			
			Tertiary (R)	Neogene (N)	Pliocene (N_2)	5	Angiosperm	Mammal	The main coal forming period
					Miocene (N_1)	24			
				Paleogene (E)	Oligocene (E_3)	37			
					Eocene (E_2)	58			
					Paleocene(E_1)	65			
	Mesozoic (Mz)		Cretaceous (K)		Sinian (K_2)			Reptile	
					Early Cretaceous (K_1)	137			
			Jurassic (J)		Late Jurassic (J_3)		Gymnosperm		
					Middle Jurassic (J_2)				
					Lias (J_1)	203			
			Triassic (T)		Late Triassic (T_3)				The crustal movement was strong at the later period. Magma was active and sea water withdraws from the mainland
					Middle Triassic (T_2)				
					Early Triassic (T_1)	251			
	Paleozoic (Pz)	Neopaleozoic (Pz^2)	Permian (P)		Late Permian (P_2)		Spore-producing plant	Amphibian	
					Early Permian (P_1)	295			
			Carboniferous (C)		Late Carboniferous (C_3)				
					Middle Carboniferous(C_2)				
					Early Carboniferous(C_1)	355			
			Devonian (D)		Late Devonian(D_3)			Fish	
					Middle Devonian (D_2)				
					Early Devonian (D_1)	408			
		Early Paleozoic (Pz^1)	Silurian (S)		Late Silurian (S_3)				The crustal movement was strong at the later period, and most places were in the shallow water environment
					Middle Silurian(S_2)				
					Llandoverian (S_1)	435			
			Ordovician (O)		Upper Ordovician (O_3)			Marine invertebrates	
					Middle Ordovician (O_2)				
					Early Ordovician (O_1)	495			
			Cambrian (Є)		Late Cambrian (Є$_3$)				
					Middle Cambrian (Є$_2$)				
					Early Cambrian (Є$_1$)	540			
Pre-Cambrian Time	Proterozoic era (Pt)		Sinian (Z)		Late Sinian (Z_2)				Sea water was transgressive widely and the crustal movement was strong at the later period
					Early Sinian (Z_1)	1 800			
							Senior algae and marine algae appeared	Lower invertebrates appeared	
	Archean (Ar)					3 200	Prokaryotes (bacteria, algae) appeared		
Early development stage of the Earth						4 000 4 600	No life		

Thinking Questions

1. What are the layers of the Earth?
2. What is endogenic geological process divided into?
3. What is exogenic geological process divided into?
4. What are time units divided into? What are time-rock units divided into?

Chapter 2

Rock Forming Minerals and Rocks

2.1 Rock Forming Minerals

Most of the chemical elements in the Earth's crust exist in the form of compounds, only a few exist in the form of elementary substance. Minerals refer to the natural elements and compounds which possess some chemical compositions and physical properties. Rocks can be formed by one or more minerals which are called rock forming minerals.

The common rock forming minerals have more than 20 categories, such as orthoclase, anorthose, black mica, white mica, augite, amphibole, talcum, kaolinite, quartz and pyrite.

2.1.1 Classification of the Minerals

Minerals found in nature exist in the certain geological environment and change with various geological processes. Minerals can form secondary minerals because of the changes of the original composition, the internal structure and properties when external conditions change to a certain extent. Therefore, rock forming minerals can be divided into primary minerals and secondary minerals according to the formation of minerals.

Primary minerals are formed by magma condensation out of the surface or on the Earth's crust, such as quartz, feldspar, mica, pyroxene, hornblende, calcite, magnetite and pyrite.

Secondary minerals are new minerals generated by original minerals which appear with chemical changes. The chemical composition and structure of secondary minerals are different from the original minerals, such as serpentine, kaolinite and sardinianite.

2.1.2 Clay Minerals

Clay minerals are aluminosilicate with schistose or chain crystal lattice and belong to secondary minerals. Clay minerals include three groups, namely kaolinite group, illite group and montmorillonite group. Chip is the most basic crystallization unit of the clay minerals, and can form unit cell gathered in a different way. The three groups consist of two types of chips, namely aluminum hydroxide chip and silica wafer chip.

Clay minerals of kaolinite group can form relatively bulky clay particle, even silt particle. The crystalline form of kaolinite group is usually elongated hexagon.

Crystal lattice of montmorillonite group minerals has the dimbibition ability and weak connecting forces, which can form flake shaped particles. The shape of crystal is an irregular circle.

The unit cells of illite group minerals are connected by potassium ions, whose connecting forces are stronger than montmorillonite group and weaker than kaolinite group. Therefore, the size of the lamellar particle is between montmorillonite group and kaolinite group.

Clay minerals have plasticity, refractory and sintering, which are the most important natural raw materials of ceramics, refractory material, cement and other industries.

2.1.3 Morphology of Minerals

1. Monomer Form

Solid minerals are divided into crystalline minerals and amorphous minerals according to their array types of internal particles (atoms, ions, molecules) in 3D space. The internal particles of crystalline minerals array regularly, therefore crystal has a certain internal structure and a geometrical shape under the conditions of appropriate growth. For example, NaCl is a cubic lattice structure in 3D space, and its geometric shape is a cube.

Rock forming minerals are the vast majority of crystalline minerals. Different crystalline minerals have different geometric shapes because of different internal

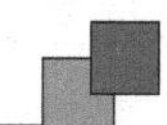

structures, such as rhombohedron calcite, flaky mica and cube pyrite or pyritohedron pyrite.

The same grains always tend to form a certain crystalline form under the same growth conditions, which is called crystal habit.

According to the growing degree of the crystal in three-dimensional space, crystal habit can be divided into three categories as follows:

① When the crystal grows mainly along one direction, it can appear columnar, acicular or threadiness, such as hornblende, pyroxene, tourmaline and so on.

② When the crystal grows mainly along two directions, it can appear clintheriform, schistose and flaky, such as mica, chlorite and so on.

③ When the crystal grows along three directions, it can appear equiaxed shape and granularity, such as garnet, peridot and so on.

Particles of amorphous minerals arrange irregularly, so that they can not form a certain geometric shape, such as opal, agate, volcanic and so on.

2. Aggregate Forms

Minerals often are identified by their aggregate forms, because crystalline minerals are rarely seen in monomer in nature, and amorphous minerals have no regular monomer form. Aggregation is formed by many monomers of the same minerals, whose form depends on the morphology of monomer and aggregate ways.

Aggregate can be divided into macroscopic crystalline mineral aggregate and cryptocrystal line or amorphous mineral aggregate that can not be identified by naked eyes according to the sizes of the mineral crystal particles.

Common aggregate forms of minerals include druse, threadiness, graininess, stalactitic form, oolith, earthy form and massive form.

Druse refers to the crystal group of the same kind of mineral in the same substrate, such as crystal clusters, calcite crystal clusters and so on.

Threadiness is formed by a lot of acicular, columnar or hair-like monomer minerals, such as asbestos, fiber gypsum and so on.

Crystals of the similar size and random arrangement can form granular aggregate, which can be divided into three types according to the particle sizes, namely coarse granular, granular and fine granula.

Stalactitic form appears in limestone caves commonly. Stalactite refers to the cystolith hangs down from the top of the cave. Stalagmite refers to the cystolith grows from the bottom of the cave. When stalactites and stalagmites are meeting, they become stone pillars.

Colloidal substance can become tuberculosis around a particle, and then tuberculosis aggregates into collection, whose shape is like a roe, such as oolitic hematite.

Earthy form refers to the aggregate with the state of the loose powder, such as kaolinite.

Massive form refers that fine minerals tightly integrated together without a certain arrangement, such as opal, massive quartz and so on.

2.1.4 Physical Properties of Minerals

The physical properties of minerals include colour, streak, gloss, transparency, hardness, cleavage, fracture, etc.

1. Colour

Minerals colours are reflection results of absorption and reflection degree of different wavelengths visible lights, which can be divided into essential colour, allochromatic colour and faussecouleur.

Essential colour is the inherent colour of the minerals, for example, pyrite is brassy, calcite is white, and olivine is olive green.

Allochromatic colour appears when mixing with impurities, which has nothing to do with minerals and can not be used to identify minerals. For example, pure quartz is colourless; it can appear milky white, purple red, green and other colours when it contains different impurities.

Faussecouleur appears when internal crack or surface oxide film of minerals refracts or scatter lights. For example, the rainbow appeared on joint surface of calcite is faussecouleur.

2. Streak

Streak refers to the colour of remaining powder when mineral scratches on the white unglazed porcelain plate. It has the vital significance to identify dark minerals because it can eliminate faussecouleur and weaken allochromatic colour. Streak of some minerals is different from its colour, for example, pyrite is yellow and has black green streak, hematite may be red, steel gray, ironish black, but its streak is always cerise .

3. Gloss

Gloss refers to the reflection ability of minerals' smooth surfaces. According to the intensity of reflected light, gloss is divided into three types, namely metallic

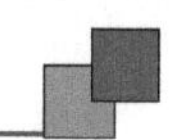

luster (such as galena and pyrite), submetallic luster (such as magnetite) and non-metallic luster (such as iceland spar).

Rock forming minerals are mostly belonging to nonmetallic luster minerals. Due to the nature of the minerals' surface or the aggregate way of minerals' collection, they can reflect different characteristics luster, such as glassy luster (on the cleavage planes of the feldspar and calcite), oily luster (on the fracture of quartz), nacreous luster (isinglass), silky luster (fiber gypsum and sericite), earthy luster (kaolinite and bauxite) and waxy luster (serpentine and talc).

4. Transparency

Transparency refers to the extent of the visible light going through the mineral, which depends on the chemical properties and crystal structures of minerals, and is also affected by the thickness of minerals. Transparencies of mineral samples should be ensured with the same thickness.

Depending on the transparency degree, minerals can be classified into three categories: transparent minerals (rock crystal and iceland spar), semi-transparent minerals (talc) and opaque minerals (pyrite, magnetite and graphite).

5. Hardness

Hardness is an important feature of mineral identification, which refers to the property to fight curving and grinding by external force. Generally, the relative hardness of minerals is determined by curving each other. The standard of hardness is made up of 10 kinds of minerals, which is called Mohs Hardness Scale (shown in Table 2.1).

Table 2.1 Mohs Hardness Scale

Nonminerals	1	2	3	4	5	6	7	8	9	10
Mineral	Talc	Gypsum	Calcite	Fluorite	Apatite	Feldspar	Quartz	Topaz	Corundum	Diamond

It's worth nothing that Mohs Hardness Scale only reflects the relative hardness orders of the minerals, not absolute hardness order. Nonminerals of common rock forming minerals are mainly between 2 and 6.5, only a few minerals (such as quartz, olivine and garnet) are greater than 6.5. Nonminerals of fingernail is 2.5, and penknife is 5.5, which are always used to identify the hardness of minerals.

6. Cleavage and Fracture

Another two important identification characters are cleavage and fracture, which affect the mechanical strength of the rocks. Cleavage planes refer to the smooth planes which form when crystalline minerals split along a certain direction under the action of hitting. If it is an uneven cross section, it is termed as fracture. Cleavage can be divided into four types, namely, eminent cleavage, perfect cleavage, moderate cleavage and imperfect cleavage.

Eminent cleavage can appear easily, and cleavage plane is large, completed and smooth. There is no fracture in these minerals, such as mica.

Cleavage plane of perfect cleavage is smooth, and minerals are easily split into laminated or small pieces, such as calcite.

Cleavage plane of moderate cleavage is not very smooth, such as hornblende.

Cleavage plane of imperfect cleavage may not exist; it often appears fracture, such as quartz and apatite.

There is no fracture when there is eminent cleavage and vice versa. For example, quartz only has fracture and no cleavage plane. The fracture of quartz is conchoidal, native copper is crenulated, and pyrite is uneven.

In addition, there are some ways to identify the minerals. For example, talc is soapy, so it can be identified by hand, and calcite can generate bubbles when it meets hydrochloric acid.

2.1.5 Characteristics of Common Rock Forming Minerals

The characteristics of common rock forming minerals are listed in Table 2.2.

Table 2.2 Characteristics of common rock forming minerals

Minerals	Chemical component	Shape	Colour	Streak	Gloss	Nonminerals	Cleavage and fracture	Relative density	Main identification characteristics	Sample
Quartz	SiO_2	Six-party prism, granular and crystal clumps	Colorless, milk white or other colour	None	Glassy luster and oily luster	7	Conchoidal fracture	2.6	Shape, hardness and horizontal stripes in crystal cylinder	
Orthoclase	$KAlSi_3O_8$	Stumpy and clintheriform	Flesh red	None	Glassy luster	6	Two moderate cleavages are orthogonal	2.6	Cleavage and colour	
Anorthose	$(Na,Ca)[AlSi_3O_8]$	Columnar and clintheriform	White and grey white	White	Glassy luster	6	Two moderate cleavages are oblique crossing with angle 86°	2.7～3.1	Colour	

(continued)

Minerals	Chemical component	Shape	Colour	Streak	Gloss	Nonminerals	Cleavage and fracture	Relative density	Main identification characteristics	Sample
White mica	$KAl_2[AlSiO_{10}](OH,F)_2$	Flaky and schistose	Colorless	None	Glassy luster and nacreous luster	2～3	One eminent cleavage	3.0～3.2	Cleavage	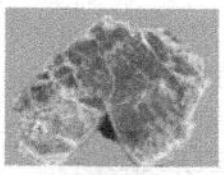
Black mica	$K(Mg,Fe)_3(AlSi_3O_{10})(OH,F)_2$	Flaky and schistose	Black and brownish black	None	Glassy luster and nacreous luster	2～3	One eminent cleavage	2.7～3.1	Cleavage and colour	
Hornblende	$(Ca,Na)2$ or $3(Mg,Fe,Al)_5[Si6(Si,Al)_2O_{22}](OH,F)_2$	Long columnar	Green black	Light green	Glassy luster	6	Two moderate cleavages are oblique crossing with angle 86° or ragged fracture	3.1～3.6	Shape and colour	
Olivine	$(Mg,Fe)_2SiO_4$	Granularity	Olive green	None	Glassy luster	6～7	Conchoidal fracture	3.3～3.5	Colour and hardness	
Calcite	$CaCO_3$	Rhomb and granularity	Colorless	None	Glassy luster	3	Three eminent cleavages	2.7	Cleavage and hardness	
Gypsum	$CaSO_4 \cdot 2H_2O$	Clintheriform and threadiness	White	White	Silky luster	2	Three cleavages, one is eminent cleavages	2.3	Cleavage and hardness	
Kaolinite	$Al_4[Si_4O_{10}](OH)_8$	Earthy and massive	White and yellow	White	Earthy luster	1	One cleavage or earthy fracture	2.5～2.6	Soft	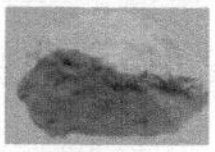
Talc	$Mg_3[Si_4O_{10}](OH)_2$	Schistose and massive	White, yellow and green	White and green	Oily luster	1	One moderate cleavage	2.7～2.8	Colour and hardness	
Garnet	$(Ca,Mg)(Al,Fe)[SiO_4]_3$	Rhombic dodecahedron, tetragonal trisoctahedron and granularity	Brownness, brownish red and green black	None	Oily luster and silky luster	6.5～7.5	No cleavage or irregular fracture	3.1～3.2	Shape, colour and hardness	
Pyrite	FeS_2	Cube and granularity	Brassy yellow	Dark green	Metallic lustre	6～6.5	Uneven fracture	4.9～5.2	Shape, colour and glossy	

2.1.6 Classification of Rocks

Rocks are aggregates of minerals, products of various geological processes and the basic materials of the lithosphere. According to their origins, rocks are classified into three groups, namely, sedimentary rocks, igneous rocks and metamorphic rocks.

2.2 Sedimentary Rocks

2.2.1 Introduction

The sedimentary rocks form an outer skin on the Earth's crust, covering 75% of the continental areas and most of the sea floor. They vary in thickness up to 10 km. Nevertheless, they only comprise about 5% of the crust.

Most sedimentary rocks are of secondary origin, in that they consist of detrital material derived by the breakdown of pre-existing rocks. Indeed, it has been variously estimated that shales and sandstones, both of mechanical derivation, account for a percentage between 75% and 95% of all sedimentary rocks. However, certain sedimentary rocks are the products of chemical or biochemical precipitation whereas others are of organic origin. Thus, the sedimentary rocks can be classified into two principal groups, namely, the clastic (detrital) (or exogenetic) type and the non-clastic (or endogenetic) type. Nevertheless, one factor that all sedimentary rocks have in common is that they are deposited, and this gives rise to their most noteworthy characteristic, that is, they are bedded or stratified.

As noted above, most sedimentary rocks are formed from the breakdown products of pre-existing rocks. Accordingly, the rate at which denudation takes place acts as a control on the rate of sedimentation, which in turn affects the character of a sediment. However, the rate of denudation is not only determined by the agents at work, that is, by weathering, or by river, marine, wind or ice action, but also by the nature of the surface. In other words, upland areas are more rapidly worn away than the lowlands. Indeed, denudation may be regarded as a cyclic process, in that it begins with or is furthered by the elevation of a land surface, and as this is gradually worn down, the rate of denudation slackens. Each cycle of erosion is accompanied by a cycle of sedimentation.

The particles of which most sedimentary rocks are composed have undergone

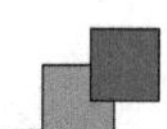

varying amounts of transportation. The amount of transport together with the agent responsible, be it water, wind or ice, play an important role in determining the character of sediment. For instance, transport over short distances usually means that the sediment is unsorted (the exception being beach sands), as does transportation by ice. With lengthier transport by water or wind, not only does the material become better sorted but also it is further reduced in size. The character of a sedimentary rock is also influenced by the type of environment in which it has been deposited, the presence of which is witnessed as ripple marks and cross bedding in sands that accumulate in shallow water.

The composition of a sedimentary rock depends partly on the composition of the parent material and the stability of its component minerals, and partly on the type of action to which the parent rock was subjected and the length of time it had to suffer such action. The least stable minerals tend to be those that are developed in environments very different from those experienced at the Earth's surface. In fact, quartz, and, to a much lesser extent, mica, are the only common detrital constituents of igneous and metamorphic rocks that are found in abundance in sediments. Most of the other minerals are ultimately broken down chemically to give rise to clay minerals. The more mature a sedimentary rock is, the more it approaches a stable end product, and very mature sediments are likely to have experienced more than one cycle of sedimentation.

The type of climatic regime in which a deposit accumulates and the rate at which this takes place also affect the stability and maturity of the resultant sedimentary product. For example, chemical decay is inhibited in arid regions so that less stable minerals are more likely to survive than in humid regions. However, even in humid regions, immature sediments may form when basins are rapidly filled with detritus derived from neighbouring mountains, while the rapid burial affords protection against the attack of subaerial agents.

In order to turn unconsolidated sediment into solid rock, it must be lithified. Lithification involves two processes, consolidation and cementation. The amount of consolidation that takes place within a sediment depends, first, on its composition and texture and, second, on the pressures acting on it, notably that due to the weight of overburden. Consolidation of sediments deposited in water also involves dewatering, that is, the expulsion of connate water from the sediments. The porosity of sediment is reduced as consolidation takes place, and, as the individual particles become more closely packed, they may even be deformed. Pressures developed during consolidation may lead to the differential solution of minerals and the authigenic growth of new ones.

Fine-grained sediments possess a higher porosity than do coarser types and, therefore, undergo a greater amount of consolidation. For instance, muds and clays may have original porosities ranging up to 80%, compared to 45% ~ 50% in sands and silts. Hence, if muds and clays could be completely consolidated (they never are), they would occupy only 20% ~ 45% of their original volume. The amount of consolidation that takes place in sands and silts varies from 15% to 25%.

Cementation involves the bonding together of sedimentary particles by the precipitation of material in the pore spaces. This reduces the porosity. The cementing material may be derived by partial intrastratal solution of grains or may be introduced into the pore spaces from an extraneous source by circulating waters. Conversely, cement may be removed from a sedimentary rock by leaching. The type of cement and, more importantly, the amount, affect the strength of a sedimentary rock. The type also influences its colour. For example, sandstones with siliceous or calcium carbonate cement are usually whitish grey, those with sideritic (iron carbonate) cement are buff coloured, whereas a red colour is indicative of hematitic (iron oxide) cement and brown of limonite (hydrated iron oxide). However, sedimentary rocks are frequently cemented by more than one material.

The matrix of a sedimentary rock refers to the fine material trapped within the pore spaces between the particles. It helps to bind the latter together.

2.2.2 Texture of Sedimentary Rocks

The texture of a sedimentary rock refers to the size, shape and arrangement of its constituent particles. Size is a property that is not easy to assess accurately, for the grains and pebbles of which clastic sediments are composed are irregular, three-dimensional objects. Direct measurement can only be applied to large individual fragments where the length of those three principal axes can be recorded. But even this rarely affords a true picture of size. Estimation of volume by displacement may provide a better measure. Because of their smallness, the size of grains of sands and silts has to be measured indirectly by sieving and sedimentation techniques, respectively. If individual particles of clay have to be measured, this can be done with the aid of an electron microscope. If a rock is strongly indurated, its disaggregation is impossible without fracturing many of the grains. In such a case, a thin section of the rock is made and size analysis is carried out with the aid of a petrological microscope, mechanical stage and micrometer.

The results of a size analysis may be represented graphically by a frequency curve or histogram. More frequently, however, they are used to draw a cumulative curve. The latter may be drawn on semi-logarithmic paper (shown in Figure 2.1).

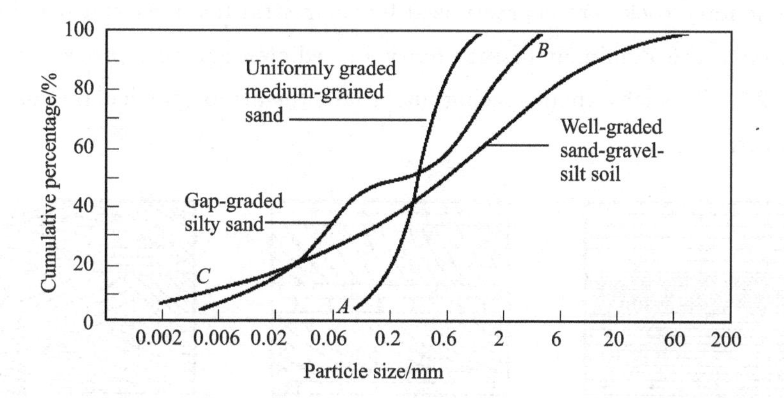

Figure 2.1 Grading curves

Various statistical parameters such as median and mean size, deviation, skewness and kurtosis can be calculated from data derived from cumulative curves. The median or mean size permits the determination of the grade of gravel, sand or silt, or their lithified equivalents. Deviation affords a measure of sorting. However, the latter can be quickly and simply estimated by visual examination of the curve in that the steeper it is, the more uniform the sorting of the sediment.

The size of the particles of a clastic sedimentary rock allows it to be placed in one of three groups that are termed rudaceous or psephitic, arenaceous or psammitic and argillaceous or pelitic.

A sedimentary rock is an aggregate of particles, and some of its characteristics depend on the position of these particles in space. The degree of grain orientation within a rock varies between perfect preferred orientation, in which all the long axes run in the same direction, and perfect random orientation, where the long axes point in all directions. The latter is found only infrequently as most aggregates possess some degree of grain orientation.

The arrangement of particles in a sedimentary rock involves the concept of packing, which refers to the spatial density of the particles in an aggregate. Packing has been defined as the mutual spatial relationship among the grains. It includes grain-to-grain contacts and the shape of the contact. The latter involves the closeness or spread of particles, that is, how much space in a given area is occupied by grains. Packing is an important property of sedimentary rocks, for it is related to their degree of consolidation, density, porosity and strength.

2.2.3 Structures of Sedimentary Rocks

Sedimentary rocks are characterized by their stratification, and bedding planes are frequently the dominant discontinuity in sedimentary rock masses (shown in Figure 2.2). As such, their spacing and character are of particular importance to the engineer.

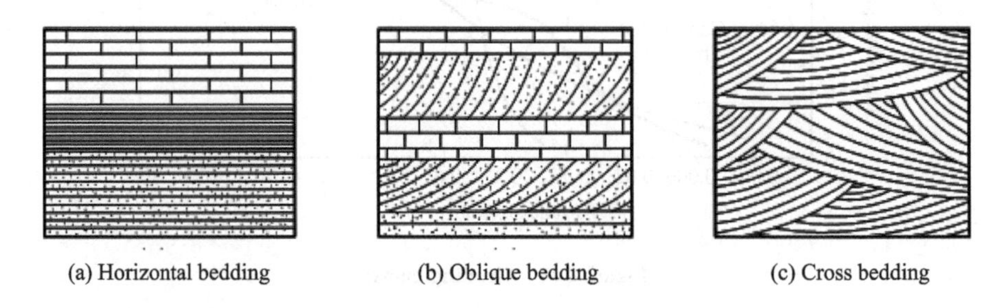

Figure 2.2 Schematic diagrams of different bedding

An individual bed may be regarded as a thickness of sediment of the same composition that was deposited under the same conditions. Lamination, on the other hand, refers to a bed of sedimentary rock that exhibits thin layers or laminae, usually a few millimeters in thickness.

Sedimentary rocks have five layer forms (shown in Figure 2.3), namely normal layer, interlayer, attenuation, wedge out and lenticle.

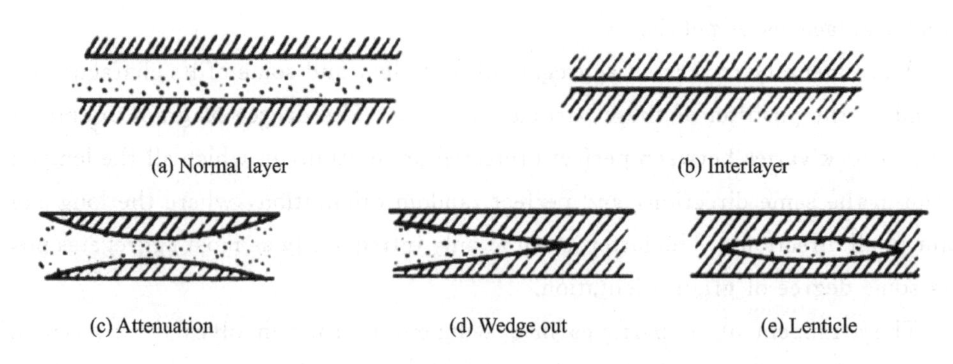

Figure 2.3 Forms of sedimentary rocks

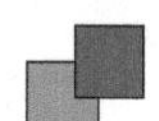

2.2.4　Sedimentary Rock Types

Sandstones (shown in Figure 2.4) consist of a loose mixture of mineral grains (quartz and feldspar) and rock fragments. Generally, they tend to be dominated by a few minerals, the chief of which is frequently quartz. The process by which sand is turned into sandstone is partly mechanical, involving grain fracturing, bending and deformation. However, chemical activity is much more important. Sandstones distribute in many areas, which is easily mined and processed. Sandstones are most widely used in building engineering, such as carved works and decoration works.

Conglomerates (shown in Figure 2.5) and breccias (shown in Figure 2.6), belong to the sedimentary clastic rocks, refer to the content of the bulky debris, whose lowest size limit is 2 mm, are greater than 50% and the content of clay is less than 25%.

Figure 2.4　Sandstone

Figure 2.5　Conglomerate

Siltstones (shown in Figure 2.7) may be massive or laminated, the individual laminae being picked out by mica and/or carbonaceous material. Micro-cross-bedding is frequently developed and the laminations may be convoluted in some siltstones. Frequently, siltstones are interbedded with shales or fine-grained sandstones, the siltstones occurring as thin ribs. The siltstones are porous, of low strength and poor stability.

Shales (shown in Figure 2.8) are the most common sedimentary rock and characterized by their lamination. Sedimentary rock of similar size range and composition, which is however not laminated, is referred to as mudstone (shown in Figure 2.9). In fact, there is no obvious distinction between shales and mudstones, one grading into the other. Shales have high water imbibition and become soft easily after water absorption.

Figure 2.6 Breccia

Figure 2.7 Siltstone

Figure 2.8 Shale

Figure 2.9 Mudstone

Limestone (shown in Figure 2.10) refers to the rock in which the carbonate fraction exceeds 50%, and over half exist as a form of calcite or aragonite ($CaCO_3$). If the carbonate material is made up mainly of dolomite ($CaCO_3$, $MgCO_3$), the rock is named dolostone (shown in Figure 2.11). Limestones and dolostones constitute 20% to 25% of the sedimentary rocks, according to Pettijohn (1975). Both are very useful building stones because of their wide distribution, uniform rock characteristics and easy mining.

Figure 2.10 Limestone

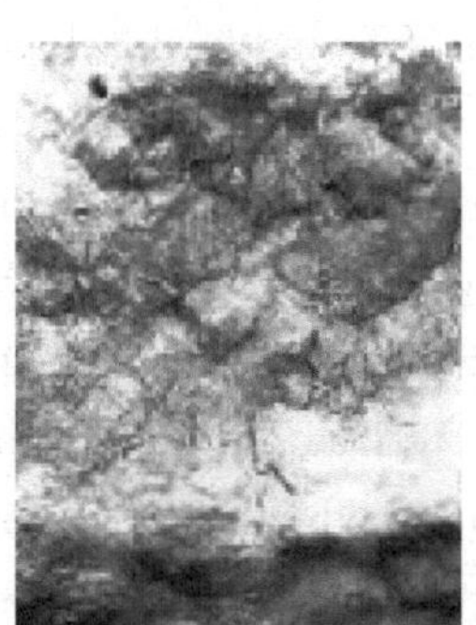

Figure 2.11 Dolostone

2.3 Igneous Rocks

2.3.1 Introduction

Igneous rocks are formed when hot molten rock material called magma solidifies. Magmas are developed when melting occurs either within or beneath the Earth's crust, that is, in the upper mantle. They comprise hot solutions of several liquid phases, the most conspicuous of which is a complex silicate phase. Thus, igneous rocks are composed principally of silicate minerals. Furthermore, of the silicate minerals, six families - the olivines [$(Mg,Fe)_2SiO_4$], the pyroxenes [e.g. augite, $(Ca, Mg, Fe, Al)_2(AL,Si)_2O_6$], the amphiboles [e.g. hornblende, $(Ca, Na,Mg,Fe,Al)_{7\ or\ 8}(Al,Si)_8O_{22}(OH)_2$], the micas [e.g. muscovite, $KAl_2(AlSi_2)_{10}(O,F)_2$, and biotite, $K(Mg,Fe)_2(AlSi_3)O_{10}(OH,F)_2$], the feldspars (e.g. orthoclase, $KAlSi_3O_8$, albite, $NaAlSi_3O_8$, and anorthite, $CaAl_2Si_2O_8$) and the silica minerals (e.g. quartz, SiO_2), are quantitatively by far the most important constituents. Figure 2.12 shows the approximate distribution of these minerals in the commonest igneous rocks.

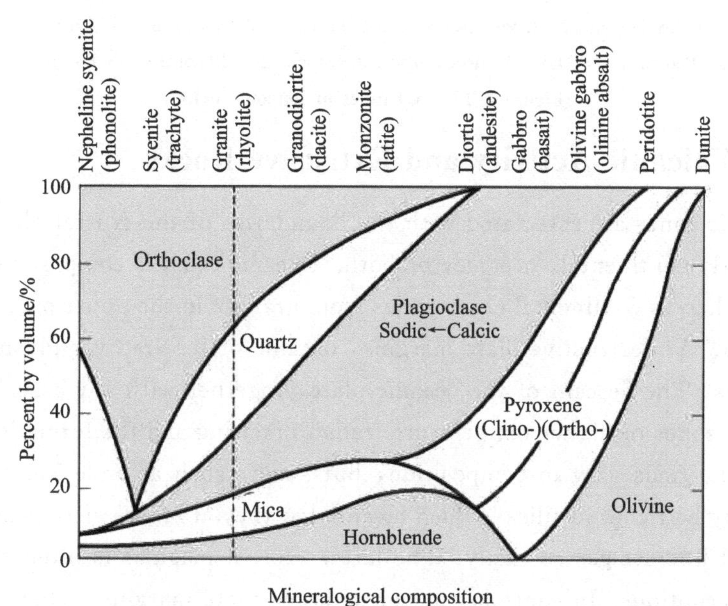

Figure 2.12 Approximate mineral compositions of the more common types of igneous rocks, e.g. granite approximately 40% orthoclase, 33% quartz, 13% plagioclase, 9% mica and 5% hornblende (plutonic types without brackets, volcanic equivalents in brackets) (By F. G. Bell)

Igneous rocks may be divided into intrusive type and extrusive type, according to their mode of occurrence. In the former type, the magma crystallizes within the Earth's crust, whereas in the latter, it solidifies at the surface, having erupted as lavas and/or pyroclasts from a volcano. The intrusions have been exposed at the surface by erosion. They have been further subdivided on the basis of their size, that is, into major (plutonic) and minor (hypabyssal) categories. The attitudes of igneous rocks are shown in Figure 2.13.

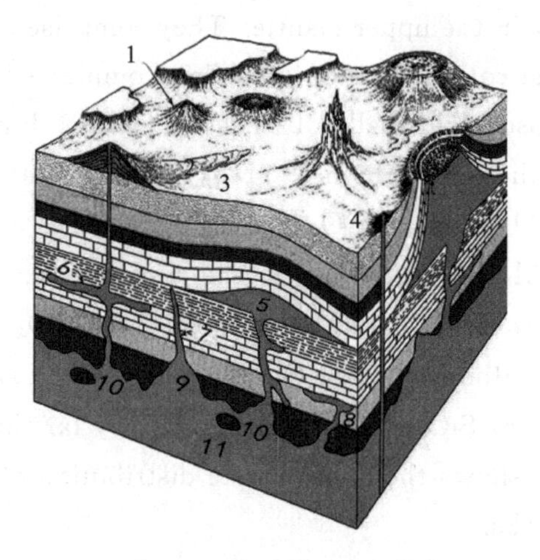

1—Volcanic cone; 2—Lava flow; 3—Lava sheet; 4—Volcanic vent; 5—Laccolite;
6—Bed rock; 7—Dyke; 8—Rock basin; 9—Stock; 10—Xenolith; 11—Batholith

Figure 2.13 Attitudes of igneous rocks

2.3.2 Volcanic Activity and Extrusive Rocks

Volcanic zones are associated with the boundaries of the crustal plates. Plates can be largely continental, oceanic, or both. Oceanic crust is composed of basaltic material, whereas continental crust varies from granitic in the upper part to basaltic in the lower. At destructive plate margins, oceanic plates are overridden by continental plates. The descent of the oceanic plate, together with any associated sediments, into zones of higher temperature leads to melting and the formation of magmas. Such magmas vary in composition, but some, such as andesitic or rhyolitic magma, may be richer in silica, which means that they are more viscous and, therefore, do not liberate gas so easily. The latter type of magmas is often responsible for violent eruptions. In contrast, at constructive plate margins, where plates are diverging, the associated volcanic activity is a consequence of magma formation in the lower crust or upper mantle. The magma is of basaltic composition, which is less viscous than andesitic or rhyolitic magma. Hence, there is relatively little

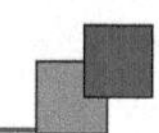

explosive activity and the associated lava flows are more mobile. However, certain volcanoes, for example, those of the Hawaiian Islands, are located in the centres of plates. Obviously, these volcanoes are unrelated to plate boundaries. They owe their origins to hot spots in the Earth's crust located above rising mantle plumes. Most volcanic materials are of basaltic composition.

Volcanic activity is a surface manifestation of a disordered state within the Earth's interior that has led to the melting of material and the consequent formation of magma. This magma travels to the surface, where it is extravasated either from a fissure or a central vent. In some cases, instead of flowing from the volcano as lava, the magma is exploded into the air by the rapid escape of the gases from within it. The fragments produced by explosive activity are known collectively as pyroclasts.

Eruptions from volcanoes are spasmodic rather than continuous. Between eruptions, activity may still be witnessed in the form of steam and vapours issuing from small vents named fumaroles or solfataras. But, in some volcanoes, even this form of surface manifestation ceases, and such a dormant state may continue for centuries. To all intents and purposes, these volcanoes appear extinct. In old age, the activity of a volcano becomes limited to emissions of the gases from fumaroles and hot water from geysers and hot springs.

Steam may account for over 90% of the gases emitted during a volcanic eruption. Other gases present include carbon dioxide, carbon monoxide, sulphur dioxide, sulphur trioxide, hydrogen sulphide, hydrogen chloride and hydrogen fluoride. Small quantities of methane, ammonia, nitrogen, hydrogen thiocyanate, carbonyl sulphide, silicon tetrafluoride, ferric chloride, aluminium chloride, ammonium chloride and argon have also been noted in volcanic gases. It has often been found that hydrogen chloride is, next to steam, the major gas produced during an eruption but that the sulphurous gases take over this role in the later stages.

At high pressures, gas is held in solution, but as the pressure falls, gas is released by the magma. The rate at which it escapes determines the explosivity of the eruption. An explosive eruption occurs when, because of its high viscosity (to a large extent, the viscosity is governed by the silica content); the magma cannot readily allow the escape of gas until the pressure that it is under is lowered sufficiently to allow this to occur. This occurs at or near the surface. The degree of explosivity is only secondarily related to the amount of gas the magma holds. Moreover, volatiles escape quietly from very fluid magmas.

Pyroclasts may consist of fragments of lava that were exploded on eruption, of fragments of pre-existing solidified lava or pyroclasts, or of fragments of country rock that, in both latter instances, have been blown from the neck of a volcano.

The size of pyroclasts varies enormously. It is dependent on the viscosity of the magma, the violence of the explosive activity, the amount of gas coming out of solution during the flight of the pyroclast, and the height to which it is thrown. The largest blocks thrown into the air may weigh over 100 tonnes, whereas the smallest consist of very fine ash that may take years to fall back to the Earth's surface. The largest pyroclasts are referred to as volcanic bombs. These consist of clots of lava or of fragments of wall rock.

2.3.3 Texture of Igneous Rocks

The degree of crystallinity is one of the most important items of texture. An igneous rock may be composed of an aggregate of crystals, of natural glass, or of crystals and glass in varying proportions. This depends on the rate of cooling and composition of the magma on the one hand and the environment under which the rock developed on the other hand. If a rock is completely composed of crystalline mineral material, it is described as holocrystalline. Most rocks are holocrystalline. Conversely, rocks that consist entirely of glassy material are referred to as holohyaline. The terms hypo-, hemi- or merocrystalline are given to rocks that are made up of intermediate proportions of crystalline and glassy materials.

2.3.4 Structure of Igneous Rocks

The structure of the igneous rocks refers to the overall characteristics of igneous rocks appearance. The common structure includes massive structure, rhyotaxitic structure, vesicular structure and amygdaloidal structure.

Massive structure refers that the mineral grains in rocks are arranged uniformly, which is the structure of the intrusive rock such as a granite and a granite porphyry.

Rhyotaxitic structure refers to the appearance characteristics of different colours and elongated stomatal bands when magma flowed over some crystallized minerals. This structure appears only in extrusive rock such as a rhyolite.

Vesicular structure refers to the pore in the Earth's surface because of low stress when magma containing volatile erupts to the Earth's surface. It is common in basic or acidic extrusive rocks such as basalt.

Amygdaloidal structure refers that the pore in extruded rocks is filled by the late mineral. It is common in the basic or neutral extrusive rocks such as basalt.

2.3.5 Igneous Rock Types

The colour index of a rock is an expression of the percentage of mafic minerals that it contains. Four categories have been distinguished:

① Leucocratic rocks, which contain less than 30% dark minerals;

② Mesocratic rocks, which contain between 30% and 60% dark minerals;

③ Melanocratic rocks, which contain between 60% and 90% dark minerals;

④ Hypermelanic rocks, which contain over 90% dark minerals.

Usually, acidic rocks are leucocratic, whereas basic and ultrabasic rocks are melanocratic and hypermelanic, respectively.

Granites (shown in Figure 2.14) and granodiorites (shown in Figure 2.15) are the most common rocks of the plutonic association and the major part of the continental crust. They are characterized by a coarse-grained, holocrystalline, granular texture. Granite is difficult to be weathered. They have beautiful colour, high hardness, wear resistance, and can be used in advanced building decoration engineering and outdoor sculpture engineering.

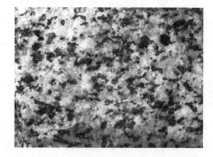

Figure 2.14 Granite

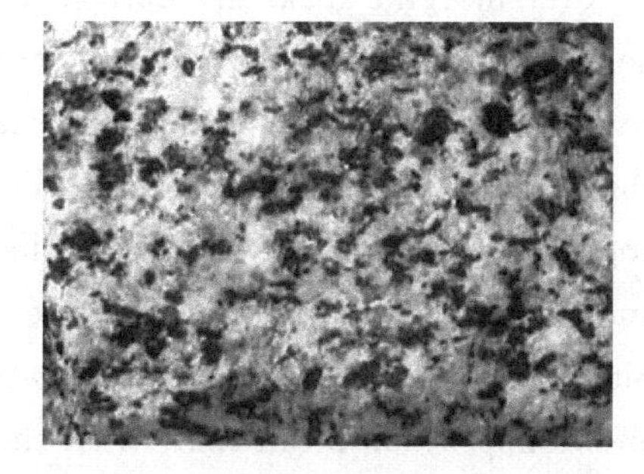

Figure 2.15 Granodiorite

Gabbros(shown in Figure 2.16) are plutonic igneous rocks with granular textures. They are dark in colour and the good road construction materials, which have high strength and anti-weathering ability.

Basalts(shown in Figure 2.17) are the extrusive equivalents of gabbros and norites, and are composed principally of calcic plagioclase and pyroxene in roughly equal amounts, or there may be an excess of plagioclase. It is by far the most important type of extrusive rock. Basalts exhibit a great variety of textures and may be holocrystalline or merocrystalline, equigranular or macro- or micro-porphyritic. Basalts are fine, solid and brittle. They have high strength, which are the good course and fine aggregates of the highway asphalt concrete pavement after broken.

Figure 2.16 Gabbro

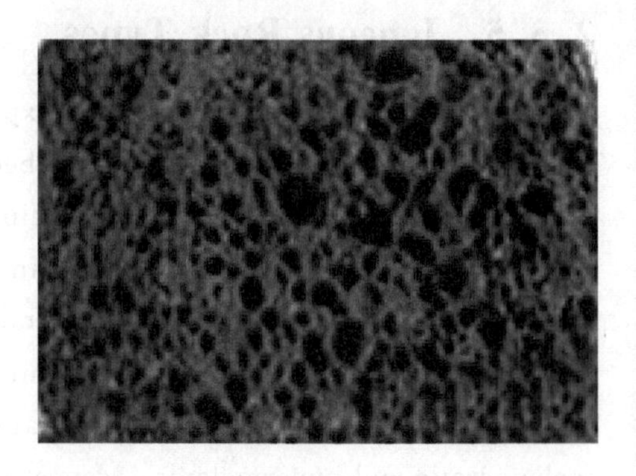

Figure 2.17 Basalt

2.4 Metamorphism and Metamorphic Rocks

2.4.1 Introduction

Metamorphic rocks are derived from pre-existing rock types and have undergone mineralogical, textural and structural changes. These changes have been brought about by changes that have taken place in the physical and chemical environments in which the rocks exist. The processes responsible for change give rise to progressive transformation in rock that takes place in the solid state. The changing conditions of temperature and/or pressure are the primary agents causing metamorphic reactions in rocks. Some minerals are stable over limited temperature-pressure conditions, which mean that when these limits are exceeded, mineralogical adjustment has to be made to establish equilibrium with the new environment.

Metamorphic reactions are influenced by the presence of fluids or gases in the pores of the rocks concerned. For instance, due to the low conductivity of rocks, pore fluids may act as a medium of heat transfer. Not only does water act as an agent of transfer in metamorphism, but it also acts as a catalyst in many chemical reactions. It is a constituent in many minerals in metamorphic rocks of low and medium grade. Grade refers to the range of temperature under which metamorphism occurs.

Two major types of metamorphism may be distinguished on the basis of geological setting. One type is of local extent, whereas the other extends over a large area. The first type refers to thermal or contact metamorphism, and the latter refers to regional metamorphism. The other type is dynamic metamorphism, which is

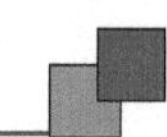

brought about by increasing stress. However, some geologists have argued that this is not a metamorphic process since it brings about deformation rather than transformation.

2.4.2 Metamorphic Textures and Structures

Most deformed metamorphic rocks possess some kind of preferred orientation. Preferred orientations may be exhibited as mesoscopic linear or planar structures that allow the rocks to be split more easily in one direction than in others. One of the most familiar examples is cleavage in slate; a similar type of structure in metamorphic rocks of higher grade is schistosity. Foliation comprises a segregation of particular minerals into inconstant bands or contiguous lenticles that exhibit a common parallel orientation.

Slaty cleavage is probably the most familiar type of preferred orientation and occurs in rocks of low metamorphic grade. It is characteristic of slates and phyllites. It is independent of bedding, which it commonly intersects at high angles; and it reflects a highly developed preferred orientation of the minerals, particularly of those belonging to the mica family.

Strain-slip cleavage occurs in fine-grained metamorphic rocks, where it may maintain a regular, though not necessarily constant, and orientation. This regularity suggests some simple relationship between the cleavage and the movement under regionally homogeneous stress in the final phase of deformation.

Harker (1939) maintained that schistosity develops in a rock when it is subjected to increased temperatures and stress that involves its reconstitution, which is brought about by localized solution of mineral material and recrystallization. In all types of metamorphisms, the growth of new crystals takes place in an attempt to minimize stress. When recrystallization occurs under conditions that include shearing stress, a directional element is imparted to the newly formed rock. Minerals are arranged in parallel layers along the direction normal to the plane of shearing stress, giving the rock its schistose character. The most important minerals responsible for the development of schistosity are those that possess an acicular, flaky or tabular habit, the micas (e.g. muscovite) being the principal family involved. The more abundant flaky and tabular minerals are in such rocks, the more pronounced is the schistosity.

Foliation in a metamorphic rock is a very conspicuous feature, consisting of parallel bands or tabular lenticles formed of contrasting mineral assemblages such as quartz-feldspar and biotite-hornblende. It is characteristic of gneisses. This parallel orientation agrees with the direction of schistosity, if any is present in nearby rocks. Foliation, therefore, would seem to be related to the same system of stress and strain responsible for the development of schistosity. However, the influence of stress becomes less at higher temperatures and so schistosity tends to disappear in rocks of high-grade metamorphism. By contrast, foliation becomes a more significant feature. What is more, minerals of flaky habit are replaced in the higher grades of metamorphism by minerals such as garnet [$Fe_3Al_2(SiO_4)_3$], kyanite (Al_2SiO_5), sillimanite (Al_2SiO_5), diopside [$Ca,Mg(Si_2O_6)$] and orthoclase.

Some metamorphic rocks are exhibited as massive structure, such as marbles, quartzites.

The metamorphic texture includes crystalloblastic texture, crush texture and palimpsest texture. Crystalloblastic structure belongs to deep metamorphic degree, which are the common structure characteristics of most metamorphic rock. Crush texture appears in the fragmentation rocks because of high pressure and temperature. Palimpsest texture is a kind of transitional type.

2.4.3 Metamorphic Types

1. Thermal or Contact Metamorphism

Thermal metamorphism occurs around igneous intrusions so that the principal factor controlling these reactions is temperature. The rate at which chemical reactions take place during thermal metamorphism is exceedingly slow and depends on the rock types and temperatures involved. Equilibrium in metamorphic rocks, however, is attained more readily at higher grades because reaction proceeds more rapidly.

2. Regional Metamorphism

Metamorphic rocks extending over hundreds or even thousands of square kilometres are found exposed in the Pre-Cambrian shields, such as those that occur in Labrador and Fennoscandia, and in the eroded roots of Fold Mountains. As a consequence, the term regional has been applied to this type of metamorphism. Regional metamorphism involves both the processes of changing temperature and stress. The principal factor is temperature, which attains a maximum of around 800 ℃ in regional metamorphism. Igneous intrusions are found within areas of regional metamorphism, but their influence is restricted. Regional metamorphism may be regarded as taking place when the confining pressures are in excess of 3×10^8 Pa. What is more, temperatures and pressures conducive to regional metamorphism must have

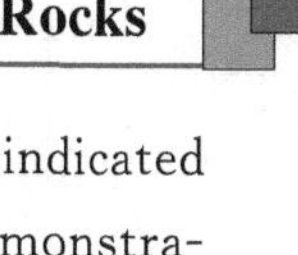

been maintained over millions of years. That temperatures rose and fell is indicated by the evidence of repeated cycles of metamorphism. These are not only demonstrated by mineralogical evidence but also by that of structures. For example, cleavage and schistosity are the results of deformation that is approximately synchronous with metamorphism but many rocks show evidence of more than one cleavage or schistosity that implies repeated deformation and metamorphism.

When sandstones are subjected to regional metamorphism, a quartzite develops that has a granoblastic (i. e. granular) texture. A micaceous sandstone or one in which there is an appreciable amount of argillaceous material, on metamorphism yields a quartz-mica schist. Metamorphism of arkoses and feldspathic sandstones leads to the recrystallization of feldspar and quartz so that granulites with a granoblastic texture are produced.

3. Dynamic Metamorphism

Dynamic metamorphism is produced on a comparatively small scale and is usually highly localized; for example, its effects may be found in association with large faults or thrusts. On a large scale, it is associated with folding, however, in the latter case; it may be difficult to distinguish between the processes and effects of dynamic metamorphism and those of low-grade regional metamorphism. What can be said is that at low temperatures, recrystallization is at a minimum and the texture of a rock is governed largely by the mechanical processes that have been operative. The processes of dynamic metamorphism include brecciation, cataclasis, granulation, mylonitization, pressure solution, partial melting and slight recrystallization.

Stress is the most important factor in dynamic metamorphism. When a body is subjected to stresses that exceed its limit of elasticity, it is permanently strained or deformed. If the stresses are equal in all directions, then the body simply undergoes a change in volume, whereas if they are directional, its shape is changed.

4. Metasomatism

Metasomatic activity involves the introduction of material into, as well as removal from, a rock mass by a hot gaseous or an aqueous medium, the resultant chemical reactions leading to mineral replacement. Thus, two types of metasomatism can be distinguished, namely, pneumatolytic (brought about by hot gases) and hydrothermal (brought about by hot solutions). Replacement occurs as a result of atomic or molecular substitution, so that there usually is little change in rock texture. The composition of the transporting medium is changing continuously because

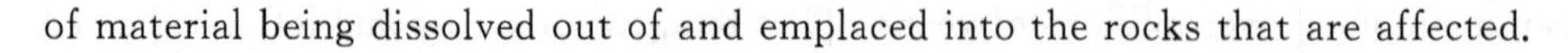

of material being dissolved out of and emplaced into the rocks that are affected.

2.4.4 Metamorphic Rock Types

Slates (shown in Figure 2.18) and marbles (shown in Figure 2.19) are the most common metamorphic rocks. Slate belongs to the schistosity rocks and dark grey or dark in colour. They are hard and can be split along the plate plane into flat slab, so that it can be widely used as building stone. Pure marbles are white in colour, so it is named "white marble" in building materials industry. Marbles containing impurities may appear other colour, such as hoary, light red, light green and dark.

Figure 2.18 Slate

Figure 2.19 Marble

Thinking Questions

1. What are the common rock forming minerals?
2. How to determine the mineral types according to Mohs Hardness Scale?
3. What are the differences among three types of rocks?
4. What are the influence factors of metamorphism?
5. Please list three kinds of building rocks and state their advantages and disadvantages.

Chapter 3

Rock Stratum and Structures

Deposition involves the build-up of material on a given surface, either as a consequence of chemical or biological growth or, far more commonly, due to mechanically broken particles being laid down on such a surface. The changes that occur during deposition are responsible for stratification, that is, the layering that characterizes sedimentary rocks. A simple interruption of deposition ordinarily does not produce stratification. The most obvious change that gives rise to stratification is in the composition of the material being deposited. Even minor changes in the type of material may lead to distinct stratification, especially if they affect the colour of the rocks concerned. Changes in grain size may also cause notable layering, and changes in other textural characteristics may help distinguish one bed from the others, as many variations in the degree of consolidation or cementation.

Since sediments are deposited, it follows that the topmost layer in any succession of strata is the youngest. Also, any particular stratum in a sequence can be dated by its position in the sequence relative to other strata. This is the Law of Superposition. This principle applies to all sedimentary rocks, of course, except those that have been overturned by folding or where older strata have been thrust over younger rocks. Where strata are overfolded, the stratigraphical succession is inverted. When fossils appeared in the beds concerned, their correct way up can be discerned. However, if fossil evidence is lacking, the correct way up of the succession may be determined from the evidence provided by the presence of "way-up" structures such as graded bedding, cross bedding and ripple marks.

3.1 Attitude of Stratum

Stratum refers to the same lithologic composition in two parallel or nearly parallel interfaces of the layer. There are two important directions associated with inclined strata, namely, dip and strike. True dip gives the maximum angle at which a bed of rock is inclined and should always be distinguished from apparent dip (shown in Figure 3.1). Inclination angle is a dip of less magnitude whose direction can run anywhere between that of the true dip and strike. The strike is the trend of inclined strata and is orientated at right angles to the true dip, which has no inclination (shown in Figure 3.1).

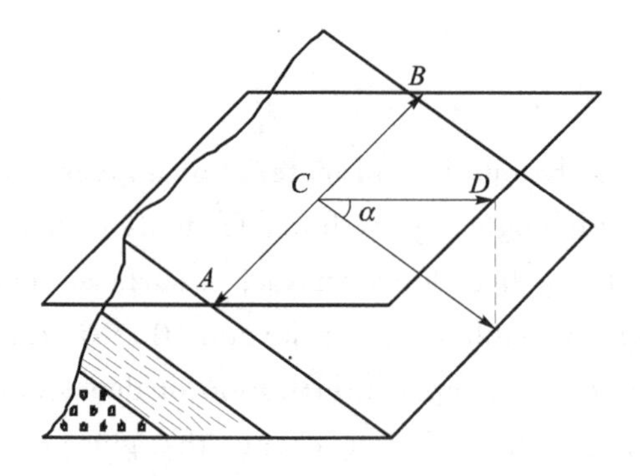

AB—Strike; *CD*—Dip; *α* —Dip angle

Figure 3.1 Illustration of dip and strike

The attitude of stratum is usually measured by geological compass and recorded by three ways as follows:

The first way is the azimuth method, which only records dip and dip angle. For example, the meaning of $230^{\circ}\angle 27^{\circ}$ is dip 230° and dip angle 27°.

The second way is the quadrant angle method, which needs to record strike, dip and dip angle.

The third way is the graphic method. The attitude of strata is marked on the geological map by using symbol. The commonly used symbols are listed as follows:

40° The long line points to strike, the short line points to dip and the number of degree points to dip angle.

The stratum is horizontal.

The stratum is vertical, and the arrow points to the newer stratum.

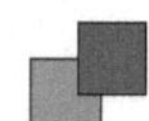

25° The stratum is inversion, and the arrow points to the dip after inversion.

3.2 Unconformities

An unconformity represents a break in the stratigraphical record and occurs when changes in the palaeogeographical conditions lead to a cessation of deposition for a period of time. Such a break may correspond to a relatively short interval of geological time or a very long one. An unconformity normally means that uplift and erosion have taken place, resulting in some previously formed strata being removed. The beds above and below the surface of unconformity are described as unconformable.

The structural relationship between unconformable units allows four types of unconformity to be distinguished. In Figure 3.2(a), stratified rocks rest upon igneous or metamorphic rocks. This type of feature has been referred to as a nonconformity (it also has been called a heterolithic unconformity). An angular unconformity is shown in Figure 3.2(b), where an angular discordance separates the two units of stratified rocks. In an angular unconformity, the lowest bed in the upper sequence of strata usually rests on beds of differing ages. This is referred to as overstep. In a disconformity, as illustrated in Figure 3.2(c), the beds lie parallel both above and below the unconformable surface, but the contact between the two units concerned is an uneven surface of erosion. When deposition is interrupted for a significant period but there is no apparent erosion of sediments or tilting or folding, then subsequently formed beds are deposited parallel to those already existing. In such a case, the interruption in sedimentation may be demonstrable only by the incompleteness of the fossil sequence. This type of unconformity has been termed a paraconformity (shown in Figure 3.2(d)).

One of the most satisfactory criteria for the recognition of unconformities is the evidence of an erosion surface between two formations. This evidence may take the form of pronounced irregularities in the surface of the unconformity. Evidence also may take the form of weathered strata beneath the unconformity, and weathering has occurred prior to the deposition of the strata above. Fossil soils provide a good example.

Stratigraphy distinguishes rock units and time units. A rock unit, such as a

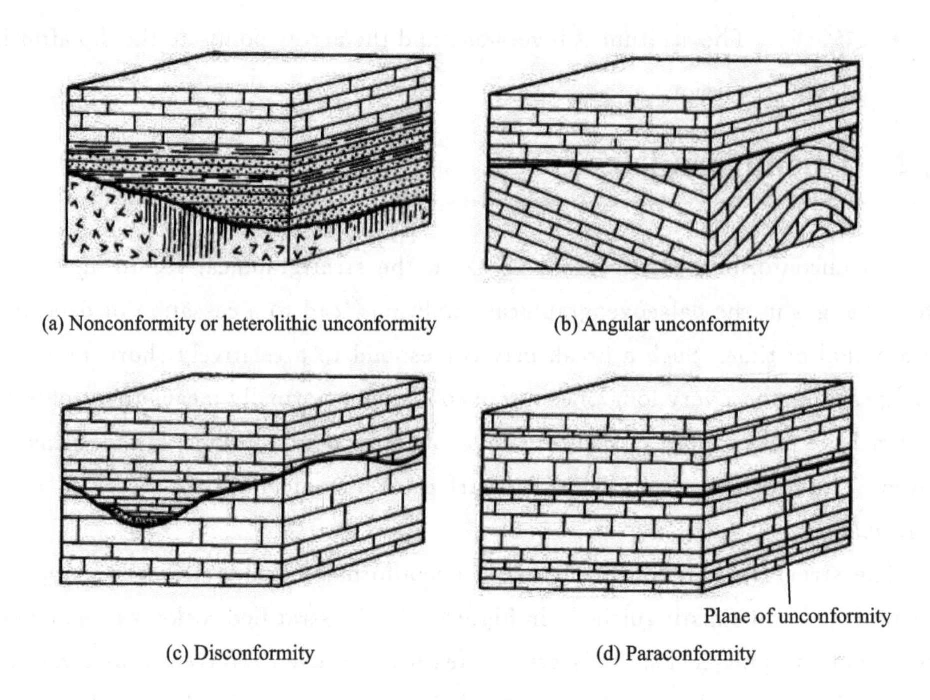

Figure 3.2 Types of unconformities

stratum or a formation, possesses a variety of physical characteristics that enable it to be recognized as such, and, hence, measured, described, mapped and analyzed. A rock unit is sometimes termed a lithostratigraphical unit.

3.3 Folds

The dip and strike of a fold can be described in the same way as those of a bedding plane. The two most important features that are produced when strata are deformed by earth movements are folds and faults, that is, the rocks are buckled or fractured, respectively. A fold is produced when a more or less planar surface is deformed to give a waved surface. On the other hand, a fault represents a surface of discontinuity along which the strata on either side have been displaced relative to each other.

Folds are wave-like in shape and vary enormously in size. Simple folds are divided into two types, that is, anticlines and synclines (shown in Figure 3.3). In the former, the beds are convex upwards, whereas in the latter, they are concave upwards.

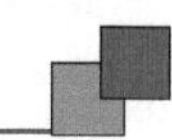

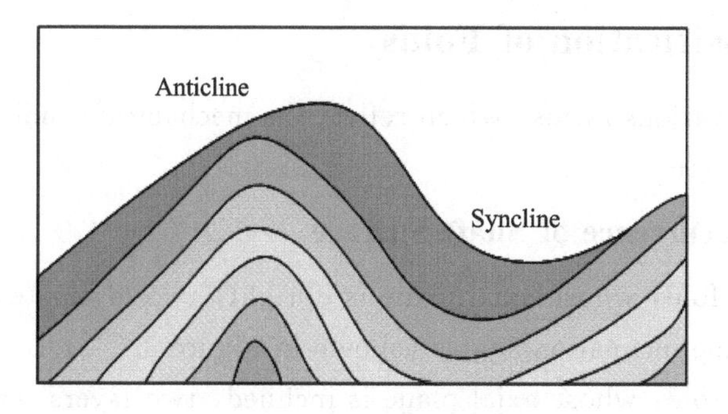

Figure 3. 3 Types of fold

3. 3. 1 Fold Elements

In order to describe the space morphological characteristics of the fold, every part of the fold is offered a certain name called fold elements (shown in Figure 3. 4).

The crestal line of an anticline is the line that joins the highest parts of the fold, whereas the trough line runs through the lowest parts of a syncline. The amplitude of a fold is defined as the vertical difference between the crest and the trough, whereas the length of a fold is the horizontal distance from crest to crest or from trough to trough.

The hinge of a fold is the line along which the greatest curvature exists and can be either straight or curved. However, the axial line is another term that has been used to describe the hinge line. The limb of a fold occurs between the hinges, all folds having two limbs. The axial plane of a fold is commonly regarded as the plane that bisects the fold and passes through the hinge line.

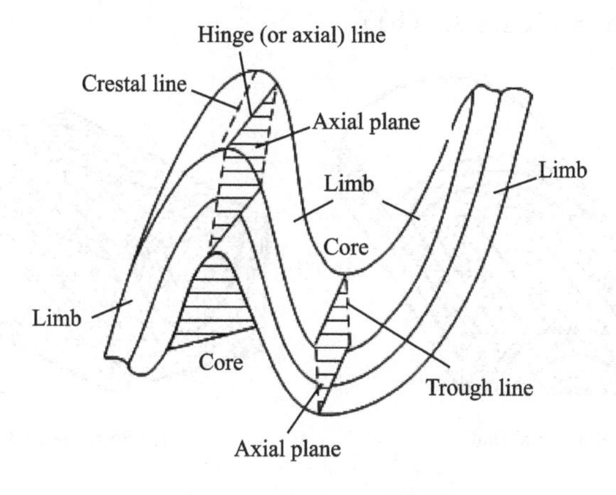

Figure 3. 4 Fold elements

3.3.2 Classification of Folds

Folds have various forms, which reflect the mechanical conditions and the causes of the formation.

1. By the Occurrence of Shaft Surface

① Upright fold, whose axial plane is upright, two layers tend to be opposite and have the same inclination angles (shown in Figure 3.5(a)).

② Inclined fold, whose axial plane is inclined, two layers tend to be opposite and have different inclination angles (shown in Figure 3.5(b)).

③ Overturned fold, whose axial plane is inclined, two layers tend to be the same, one of the limbs is inversion layer (shown in Figure 3.5(c)).

④ Recumbent fold, whose axial plane is nearly horizontal, two layers tend to be horizontal, one of the limbs is inversion layer (shown in Figure 3.5(d)).

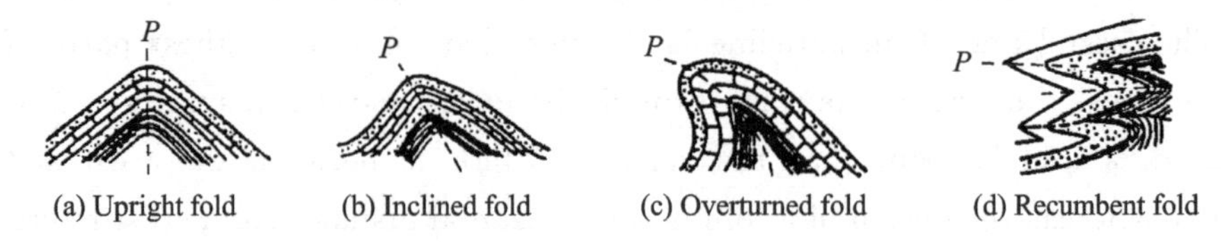

Figure 3.5 Classification by the occurrence of shaft surface

2. By the Occurrence of Hinge

① Horizontal fold, whose hinge is nearly horizontal, strike of two layers is parallel and opening (shown in Figure 3.6(a)).

② Plunging fold, whose hinge is inclined, strike of two layers is not parallel and closed (shown in Figure 3.6(b)).

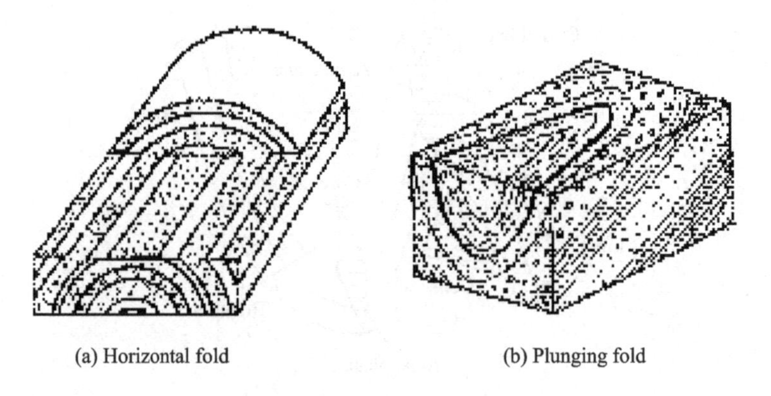

Figure 3.6 Classification by the occurrence of hinge

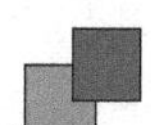

3.3.3 Types of Fold Structures

1. Anticlinorium and Synclinorium

Folds often appear in the space in the combination forms of multiple consecutive anticlines and synclines. Anticlinorium (shown in Figure 3.7(a)) refers to a huge anticline, which is composed of a series of continuous curved rocks. Synclinorium (shown in Figure 3.7(b)) refers to a huge syncline, which is composed of a series of continuous curved rocks. The two forms often appear in the area with intense tectonic movement.

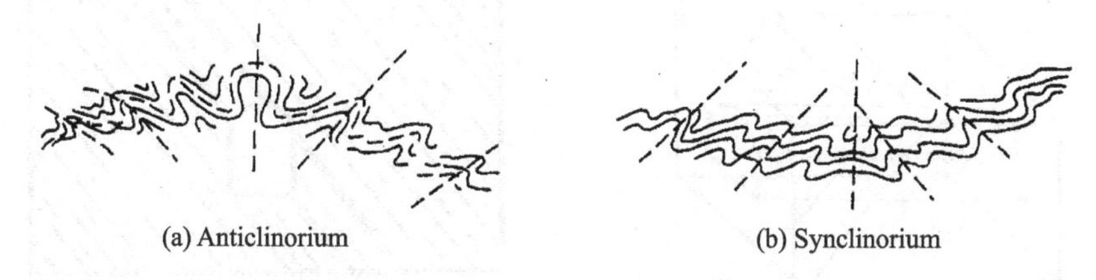

(a) Anticlinorium (b) Synclinorium

Figure 3.7 Anticlinorium and Synclinorium

2. Wide Spaced Anticlines and Wide Spaced Synclines

Fold structures can be divided into wide spaced anticlines and wide spaced synclines, which are composed of a series of continuous folds, whose axis lines are parallel in the planes. The two forms often appear in the area with relatively moderate tectonic movement. When anticline is narrow and syncline is wide, it is called wide spaced anticlines (shown in Figure 3.8(a)). When anticline is wide and syncline is narrow, it is called wide spaced synclines (shown in Figure 3.8(b)).

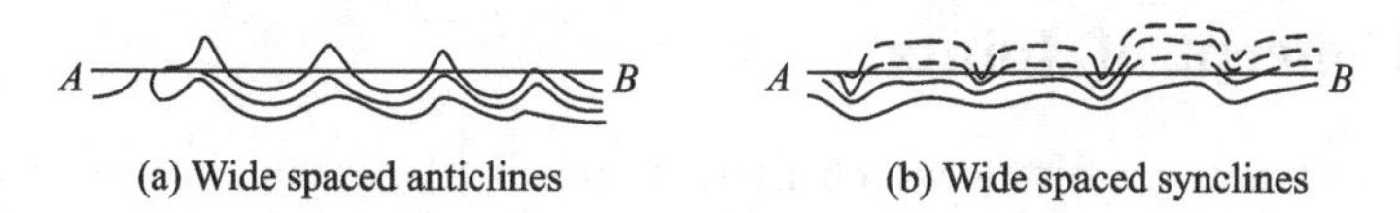

(a) Wide spaced anticlines (b) Wide spaced synclines

Figure 3.8 Wide spaced anticlines and wide spaced synclines

3.3.4 Relationship between Stratum, Fold and the Stability of Tunnel

The stability of tunnels is influenced by geological structure; therefore the construction of the tunnel should follow the principles in below.

① Tunnels should locate in hard, intact and thick layers when they go through horizontal stratum, such as limestone layer or sandstone layer.

② The top of tunnels should set up in hard rocks when they go through soft and hard rocks for avoiding collapse.

③ The top of tunnels often occurs bedding landslides in soft rock when they vertically cross soft and hard rocks. Supporting in time is the best way to make sure the stability of tunnels (shown in Figure 3. 9).

④ Tunnels easily collapse and slide when the axial lines of tunnels are parallel to the strike of the inclined rocks. Reinforcement treatment should be used to make the stability of the side wall (shown in Figure 3. 10).

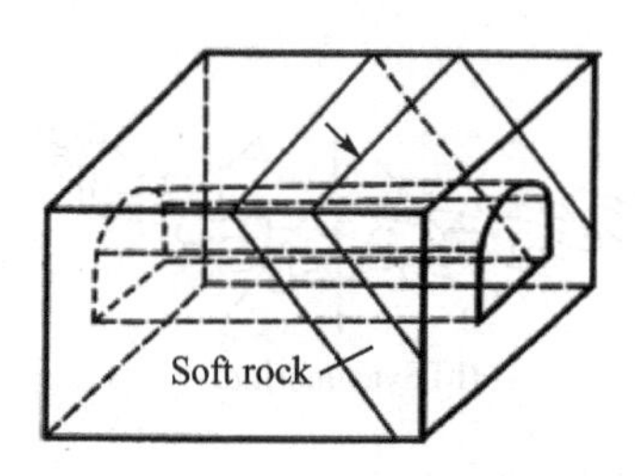

Figure 3. 9 Tunnel goes through soft and hardsocks

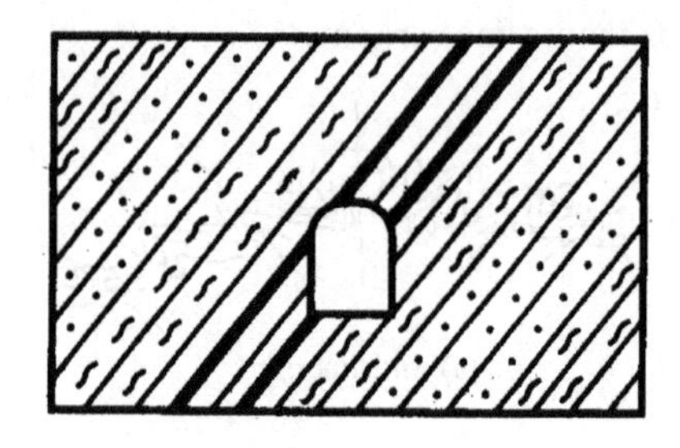

Figure 3. 10 Tunnel goes through tilted stratum

⑤ Tunnels should locate in the limb of folds and avoid core parts. Stratum in core is fractured, so groundwater can permeate into the tunnel.

3. 4 Joints

3. 4. 1 Concept of Joints

Joints are fractures along which little or no displacement has occurred and are present within all types of rocks. At the ground surface, joints may open as a consequence of denudation, especially weathering, or the dissipation of residual stress.

A group of joints that run parallel to each other are termed a joint set (shown in Figure 3. 11), whereas two or more joint sets that intersect at a more or less constant angle are referred to as a joint system. If one set of joints is dominant, then the joints are known as primary joints, and the other set or sets of joints are termed secondary. If joints are planar and parallel or sub-parallel, they are described as systematic; conversely, when their orientation is irregular, they are termed non-systematic.

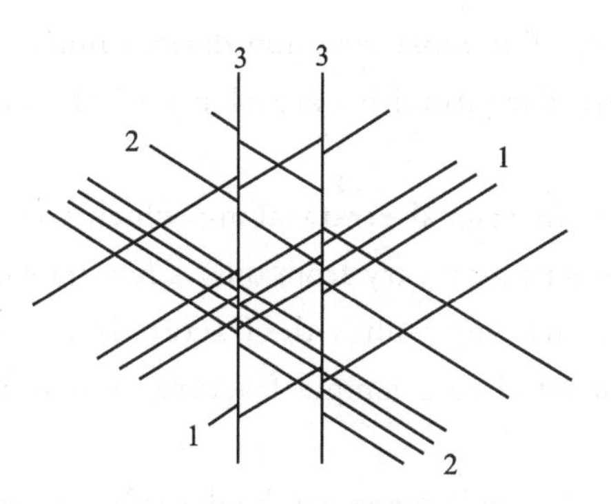

Figure 3.11 Joints

3.4.2 Classification of Joints

Joints can be divided, on the basis of size, into master joints that penetrate several rock horizons and persist for hundreds of metres, and major joints that are smaller joints but which are still well-defined structures, and minor joints that do not transcend bedding planes. Lastly, minute fractures occasionally occur in finely bedded sediments, and such micro-joints may be only a few millimetres in size.

Joints are formed through failure of rock masses in tension, in shear or through some combination of both. Rupture surfaces formed by extension tend to be clean and rough with little detritus. They tend to follow minor lithological variations. Simple surfaces of shearing are generally smooth and contain considerable detritus. They are unaffected by local lithological changes.

Joints also are formed in other ways. For example, joints develop within igneous rocks when they cool down, and in wet sediments when they dry out. The most familiar of these are the columnar joints in lava flows, sills and some dykes. The cross joints, longitudinal joints, diagonal joints and flat-lying joints associated with large granitic intrusions have been referred to in Chapter 2.

3.5 Faults

3.5.1 Concept and Elements of Faults

Discontinuity represents a plane of weakness within a rock mass across which the rock material is structurally discontinuous. Although discontinuities are not necessarily planes of separation, most in fact are and they possess little or no tensile strength. Discontinuities vary in size from small fissures on the one hand to huge faults on the other hand. The most common discontinuities are joints and bedding planes. Other important discontinuities are planes of cleavage and schistosity, fissures and faults.

Faults are fractures in crustal strata along which rocks have been displaced. The amount of displacement may vary from only a few tens of millimetres to several hundreds of kilometres. In many faults, the fracture is a clean break; in others, the displacement is not restricted to a simple fracture, but is developed throughout a fault zone.

The dip and strike of a fault plane can be described in the same way as those of a bedding plane. The angle of hade is the angle enclosed between the fault plane and the vertical. The hanging wall of a fault refers to the upper rock surface along which displacement has occurred, whereas the foot wall refers to the rock below the fault plane. The vertical shift along a fault plane is called the throw, and the term heave refers to the horizontal displacement. Where the displacement along a fault has been vertical, then the terms downthrow and upthrow refer to the relative movement of strata on opposite sides of the fault plane. The elements of faults are shown in Figure 3. 12.

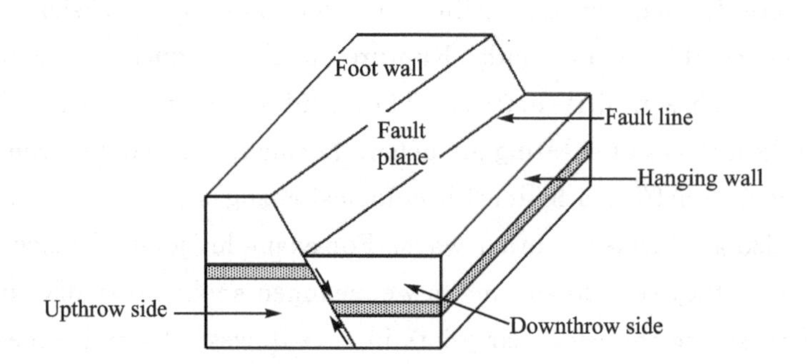

Figure 3. 12 Elements of faults

3. 5. 2 Classification of Faults

The classifications of faults can be based on the direction in which movement takes place along the fault plane, on the relative movement of the hanging and foot walls, on the attitude of the fault in relation to the strata involved and on the fault pattern. If the direction of slippage along the fault plane is used to distinguish between faults, then three types may be recognized, namely, dip-slip faults, strike-slip faults and oblique-slip faults. In a dip-slip fault, the slippage occurs along the dip of the fault. In a strike-slip fault, the slippage takes place along the strike . In

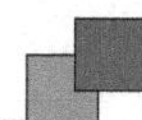

an oblique-slip fault, movement occurs diagonally across the fault plane.

When the relative movement of the hanging walls and the foot walls is used as a basis of classification, then normal faults, reverse faults and wrench faults can be recognized. A normal fault is characterized by the occurrence of the hanging wall on the downthrow side, whereas the foot wall occupies the downthrow side in a reverse fault (shown in Figure 3. 13). Reverse faults involves a vertical duplication of strata, unlike normal faults where the displacement gives rise to a region of barren ground (shown in Figure 3. 14). In a wrench fault, neither the foot walls nor the hanging walls have moved up or down in relation to one another (shown in Figure 3. 15).

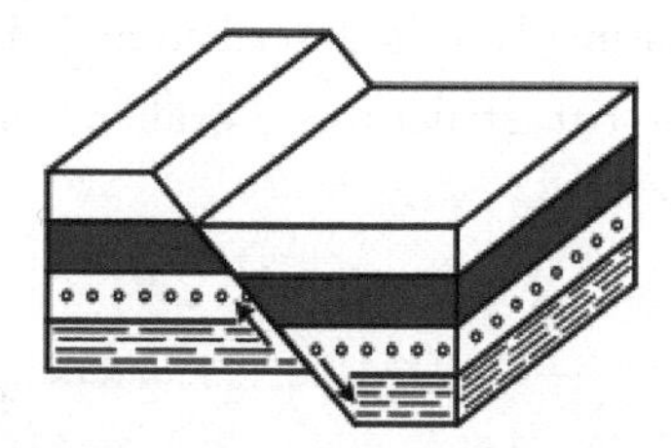

(a) Hanging wall moves down along dip direction

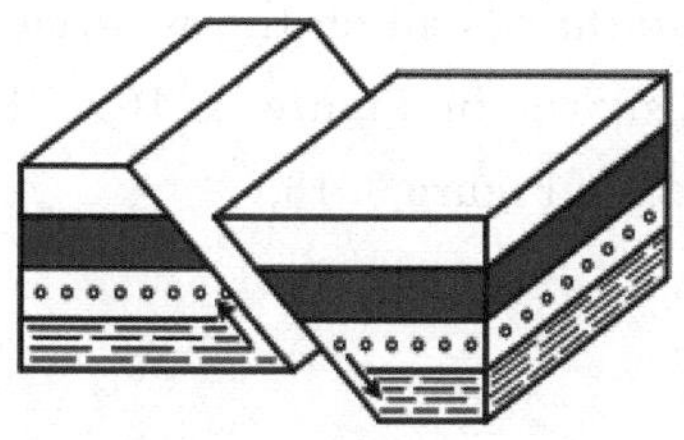

(b) Hanging wall moves down along strike direction

Figure 3. 13 Normal fault

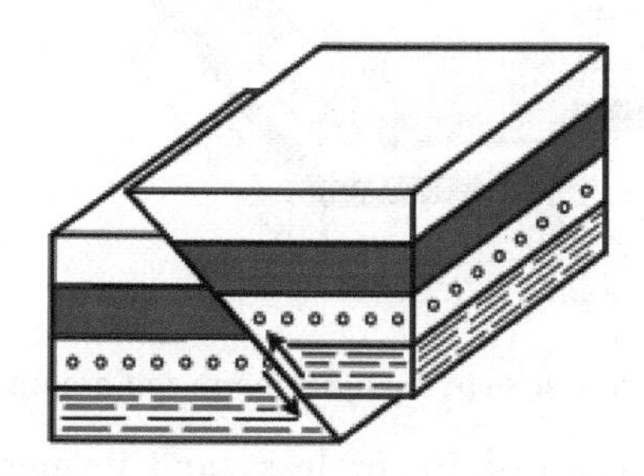

(a) Hanging wall moves up along dip direction

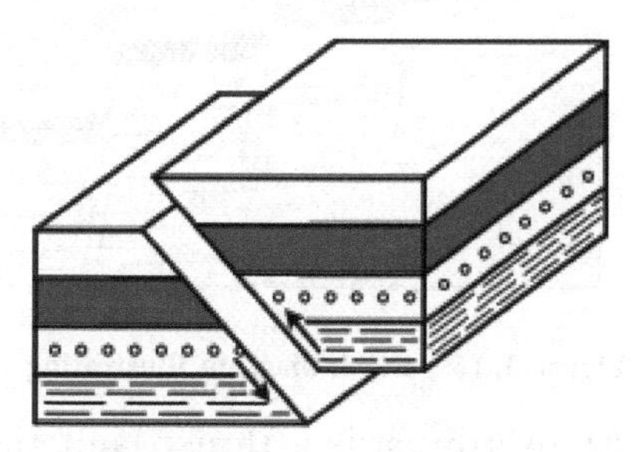

(b) Hanging wall moves up along strike direction

Figure 3. 14 Reverse fault

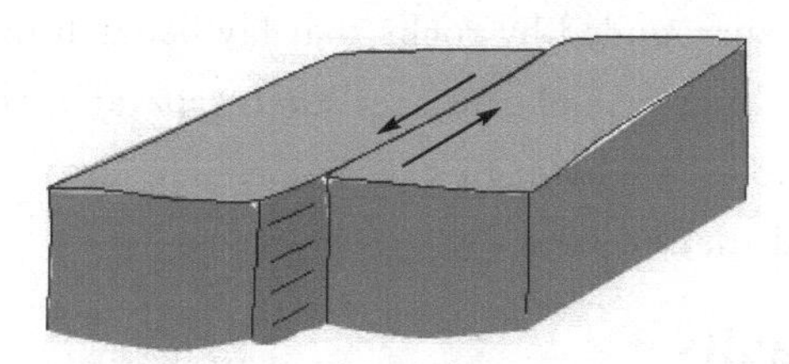

Figure 3. 15 Wrench fault arrows show the directions of relative displacement

Considering the attitude of the fault to the strata involved, strike faults, dip (or cross) faults and oblique faults can be recognized. A strike fault is one that trends parallel to the beds it displaces, a dip (or cross) fault is one that follows the inclination of the strata, and an oblique fault runs at angle with the strike of the rocks it intersects.

Normal faults range in linear extension up to, occasionally, a few hundreds of kilometres in length. Generally, the longer faults do not form single fractures throughout their entirety but consist of a series of fault zones. The net slip on such faults may be totally over a thousand meters. Normal faults are commonly quite straight in outline but sometimes they may be sinuous or irregular with abrupt changes in strike. When a series of normal faults run parallel to one another with their downthrows all on the same side, the area involved is described as being step faulted (shown in Figure 3. 16). Horsts and rift structures (graben) are also illustrated in Figure 3. 16.

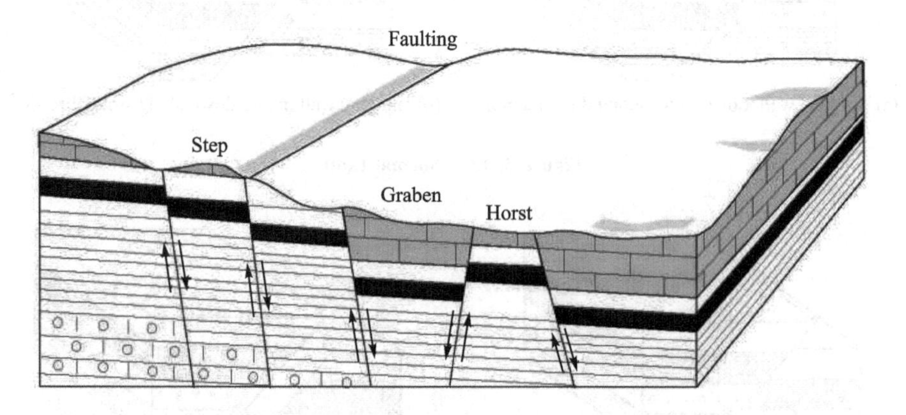

Figure 3. 16 Block diagram illustrating the structures of step, faulting, horst and graben

An overthrust is a thrust fault that has a dip of 10° or less, and its net slip measures several kilometres. Overthrusts may be folded or even overturned. As a consequence, when they are subsequently eroded, remnants of the overthrust rocks may be left as outliers surrounded by rocks that lay beneath the thrust. These remnant areas are termed klippe, and the area that separates them from the parent overthrust is referred to as a fenster or a window. The area that occurs in front of the overthrust is called the foreland.

3. 5. 3 Active Faults

Active faults are termed that faults are still in activity nowadays or maybe continue to motion in the near future. Engineering active faults refer to the active

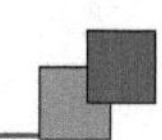

faults which may affect or do harm on the safety of civil infrastructure lifetime, usually from 50 years to 100 years. Shear moving or earthquake aroused by active faults may destroy buildings.

Active faults can be divided into creep type faults and stick slip faults according to the nature of the activity. The former terms that rocks on both sides of fault slide slowly and keep constant mutual dislocation. Creep type faults have low rock mass strength and weak fillings in fault zones so that they cannot accumulate large strain energy. Stick slip faults slide suddenly in the way of earthquake. Stick slip faults have high rock mass strength, whose walls stick together so that they cannot slip. They can accumulate large strain energy to lead to an earthquake after the stress reaches the ultimate strength of surrounding rocks.

Active faults have two characteristics, namely, inheritance and periodicity. Most active faults may displace along the old faults, that is called inheritance of active fault, which includes the same activity type and motion direction. The newer the age of active faults is, the stronger the inheritance is. The movement process of active fault often experiences activity, calm and activity process, the repetition cycle is termed the cycles of active fault.

The follow principles should be noticed when civil infrastructures are in fracture areas.

① In the fault zones, the appropriate construction types and structural measures should be adopted.

② The site of building, especially the important permanent structures such as dams and nuclear power plants, should keep away from active faults.

③ When linear engineering such as railways, highways and tunnels cross faults, they should intersect fault with a large angle.

④ If some major projects must be built in fault zones, they should locate in a relatively stable land area in the unstable zones, which is known as the "safety island". At the same time, civil engineering structures should locate in the foot wall of the fault and several kilometers away from the main fracture zone.

3.6 Geological Map

3.6.1 Introduction

Geological maps refer to the drawings that can reflect geological phenomena and geological conditions of some regions by using legend, symbols and colour.

They are geological information data from field measuring and the basis of engineering constructions.

One of the important ways by which the geologist can be of service is by producing maps to aid the engineer, planner and others who are concerned with the development of land. A variety of maps can be produced from engineering geomorphological, environmental geological and engineering geological to geotechnical maps (Anon, 1972). The distinction between these different types of maps is not always clear cut. Be that as it may, maps represent a means of storing and transmitting information, in particular, of conveying specific information about the spatial distribution of given factors or conditions. In addition, a map represents a simplified model of the facts, and the complexity of various geological factors can never be portrayed in its entirety. The amount of simplification required is governed principally by the purpose and scale of the map, the relative importance of particular geological factors or relationships, and the accuracy of the data and on the techniques of representation employed.

The purpose of engineering geomorphological maps is to portray the surface form and the nature and properties of the materials of which the surface is composed, and to indicate the type and magnitude of the processes in operation. Surface form and pattern of geomorphological processes often influence the choice of a site. Hence, geomorphological maps give a rapid appreciation of the nature of the ground and thereby help the design of more detailed investigations, as well as focusing attention on problem areas. Such maps recognize landforms along with their delimitation in terms of size and shape. Engineering geomorphological maps therefore should show how surface expression will influence an engineering project and should provide an indication of the general environmental relationship of the site concerned. If engineers are to obtain maximum advantage from a geomorphological survey, then derivative maps should be compiled from the geomorphological sheets. Such derivative maps generally are concerned with some aspect of ground conditions, such as landslip areas or areas prone to flooding or over which sand dunes migrate.

The principal object during an engineering geomorphological survey is the classification of every component of the land surface in relation to its origin, present evolution and likely material properties. In other words, a survey should identify the general characteristics of the terrain of an area, thereby providing a basis for

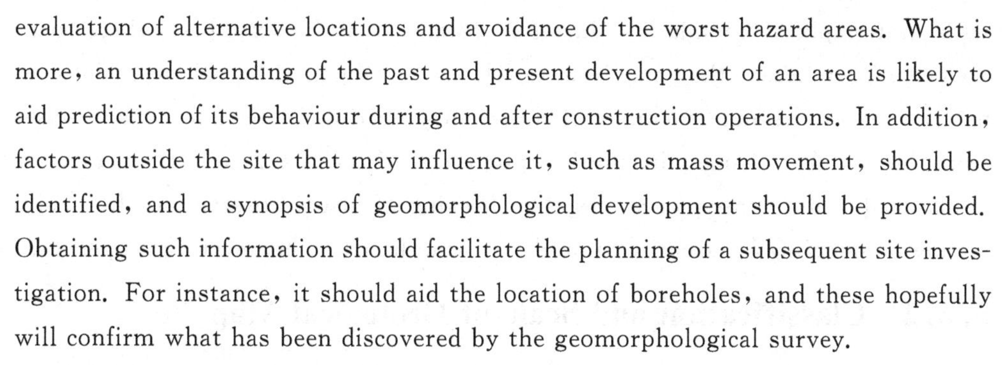

evaluation of alternative locations and avoidance of the worst hazard areas. What is more, an understanding of the past and present development of an area is likely to aid prediction of its behaviour during and after construction operations. In addition, factors outside the site that may influence it, such as mass movement, should be identified, and a synopsis of geomorphological development should be provided. Obtaining such information should facilitate the planning of a subsequent site investigation. For instance, it should aid the location of boreholes, and these hopefully will confirm what has been discovered by the geomorphological survey.

Engineering geological maps and plans provide engineers and planners with information that will assist them in the planning of land use and in the location and construction of engineering structures of all types. Such maps usually are produced on the scale of 1∶10 000 or smaller, whereas engineering geological plans, being produced for a particular engineering purpose, have a larger scale. Engineering geological maps may serve a special purpose or a multipurpose (Anon, 1976). Special purpose maps provide information on one specific aspect of engineering geology. For example, the engineering geological conditions at a dam site or along a routeway or for zoning for land use in urban development. Multipurpose maps cover various aspects of engineering geology.

Engineering geological maps should be accompanied by cross sections, and explanatory texts and legends. Detailed engineering geological information can be given, in tabular form, on the reverse side of the map. For example, a table of rock and soil characteristics summarizing the various rock and soil groups, listing their mode of occurrence, their thickness, their structure and their hydrogeological and geotechnical properties may be provided. More than one map of an area may be required to record all the information that has been collected during a survey. In such instances a series of overlays or an atlas of maps can be produced. Preparation of a series of engineering geological maps can reduce the amount of effort involved in the preliminary stages of a site investigation, and may indeed allow site investigations to be designed for the most economical confirmation of the ground conditions.

Geotechnical maps and plans indicate the distribution of units, defined in terms of engineering properties. For instance, they can be produced in terms of index properties, rock quality or grade of weathering. A plan for a foundation could be made in terms of design parameters. The unit boundaries then are drawn for changes in the particular property. Frequently, the boundaries of such units coincide with stratigraphical boundaries. In other instances, as for example, where rocks are deeply weathered, they may bear no relation to geological boundaries. Unfortunately, one of the fundamental difficulties in preparing geotechnical maps arises from the fact that changes in physical properties of rocks and soils frequently are gradational. As a consequence, regular checking of visual observations by in-situ testing or sampling is essential to produce a map based on engineering properties.

3. 6. 2 Classification and Scale of Geological Map

Geological plan uses a variety of legends to describe all kinds of geological data, such as landform, strata, geologic structure, and natural geological and hydrogeological conditions on the plane. Geological profile refers to the map which can reflect deep strata and geological structure of datum. Comprehensive strata log diagramrefers to the columnar section which can reflect stratigraphic sequence, thickness and lithology characteristics by using a certain scale and legend, and regional geology history.

Scale reflects the precision indicators of map. The larger the scale of map is, the higher the precision is, which can reflect the more detailed and accurate contents.

According to the scale of geologic map, geological map can be divided into small-scale geologic map (1:200 000～1:1 000 000), medium-scale geologic map (1:50 000～1:100 000) and large-scale geologic map (1:1 000～1:25 000).

3. 6. 3 Representation of Geological Structures

1. Stratum

Topographic contour refers to the closed curve of all adjacent points that has the same elevation. On the geological map, the boundary line of horizontal stratum is parallel or overlapping to the topographic contour (shown in Figure 3. 17, dotted line represents the topographic contour, black belt represents stratum). The boundary line of tilted strata on the geological map intersects with topographic contour as "V" or "U" symbol (shown in Figure 3. 18). The boundary line of upright rock is not affected by terrain contour and extends in line along the strike (shown in Figure 3. 19).

Stratum also can be identified by legends, such as symbol "⅄", long line of which represents strike and short line represents inclination.

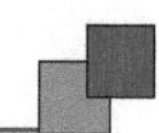

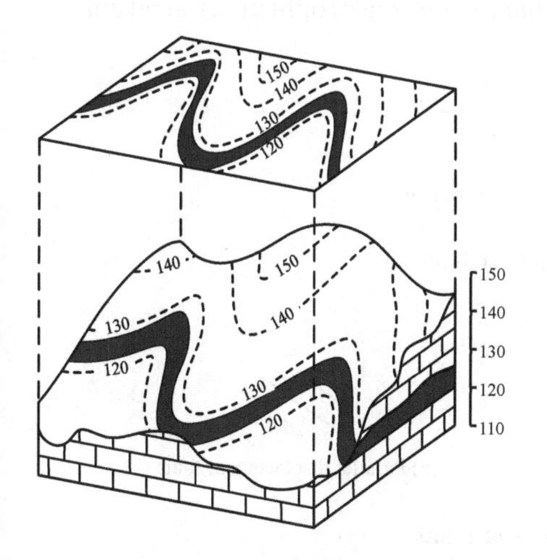

Figure 3.17 Horizontal stratums

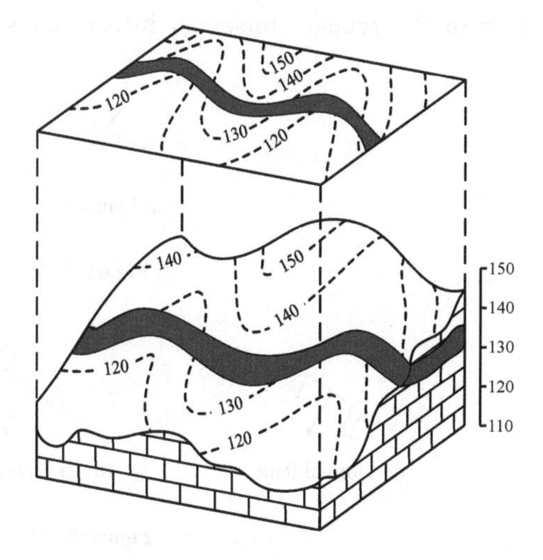

Figure 3.18 Tilted stratums

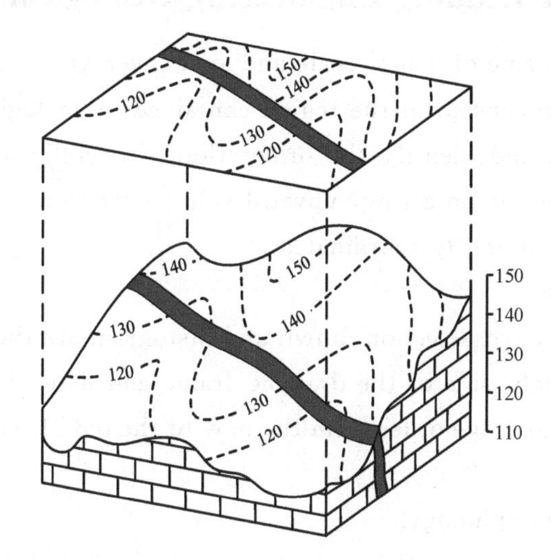

Figure 3.19 Upright stratums

2. Folds

Generally, folds can be recognized according to legends (shown in Figure 3.20). If there are no legend symbols, folds should be confirmed according to the symmetrical distributed relationship of the new and old rocks.

3. Faults

Generally, faults can be recognized according to legends (shown in Figure 3.21). If there are no legend symbols, faults should be confirmed according to the distribution of the repeat, missing, interrupt, size change and diastrophism of stratum.

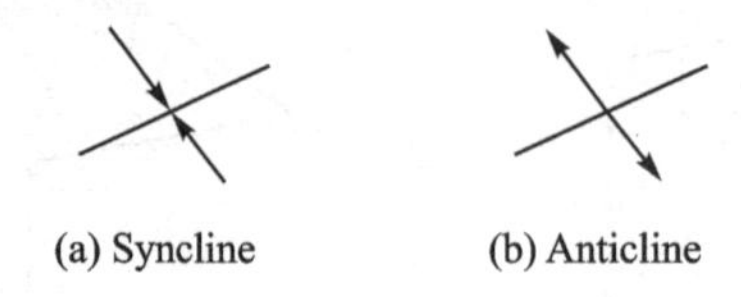

Figure 3.20 Legends of folds

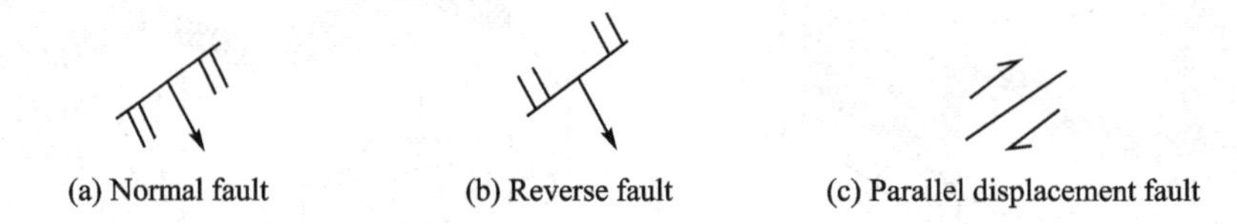

Figure 3.21 Legends of faults

3.6.4 Steps of Reading Engineering Geological Map

① Look at the name of maps, scale and area coverage.

Establish overall concept of the region contained in geological map, know the location of the map and identify the orientation. Generally, the arrow refers to north is basis. If there is no arrow, upward side points to north. Orientation also can be confirmed according to coordinates.

② Read legends.

Legends in plan, cross-section drawn and histogram are the same. In general, legends are in the right side of the drawing frame and listed from top to bottom according to the geologicage order from the new to the old. It should be paid attention to the lacuna.

③ Analyze geomorphology.

Understand mountains and rivers situation, the ups and downs of the terrain and physiognomy morphological character according to the features of the river system and topographic contour.

④ Understand strata distribution and lithology.

According to the illustrations, identify strata distribution, occurrence, the relationship between the new and old as well as the relationship with the terrain.

⑤ Identify the illustrations.

Identify the illustrations of geological structures, such as faults and their size, folds and their types.

⑥ Identify magmatic rock.

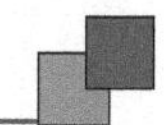

If there is magmatic rock along the exposed area, the era of magmatic activity, sequence of invasion or eruption should be made clear and then the occurrence is determined according to the output and morphological characteristics of magmatic body.

⑦ Evaluate stability.

The stability of building field should be evaluated according to the geological conditions.

Thinking Questions

1. What are the main types of geological structure?
2. What is an anticline? What is a syncline? Please list the elements.
3. Please illustrate the influence of the fold on the stability of tunnel.
4. What is a fault? What kinds can it be divided into?
5. What is an active fault? What kinds can it be divided into?
6. Please list the characteristics of an active fault.
7. Please list the principles of the civil engineerings in fault zones.

Chapter 4

Weathering, River and Groundwater

4.1 Weathering

Landmasses are continuously worn away or denuded by weathering and erosion. The agents of erosion are the sea, rivers, wind and ice. The detrital products resulting from denudation are transported by water, wind, ice or the action of gravity, and are ultimately deposited. In this manner, the surface features of the Earth are gradually, but constantly, changing.

The process of weathering represents an adjustment of the minerals of which a rock is composed to the conditions prevailing on the surface of the Earth. As such, weathering of rocks is brought about by physical disintegration, chemical decomposition and biological activity. It weakens the rock fabric and exaggerates any structural weaknesses, all of which further aid the breakdown processes. A rock may become more friable as a result of the development of fractures both between and within mineral grains. The agents of weathering, unlike those of erosion, do not themselves provide for the transportation of debris from the surface of a rock mass. Therefore, unless the rock waste is otherwise removed, it eventually acts as a protective cover, preventing further weathering. If weathering is continuous, fresh rock exposures must be constantly revealed, which means that the weathered debris must be removed by the action of gravity, running water, wind or moving ice.

Weathering also is controlled by the presence of discontinuities in that they provide access into a rock mass for the agents of weathering. Some of the earliest

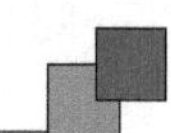

effects of weathering are seen along discontinuity surfaces. Weathering then proceeds inwards so that the rock mass may develop a marked heterogeneity with corestone of relatively unweathered material within a highly weathered matrix. Ultimately, the entire rock mass can be reduced to a residual soil. Discontinuities in carbonate rock masses are enlarged by dissolution, leading to the development of sinkholes and cavities within the rock mass.

The rate at which weathering proceeds depends not only on the vigour of the weathering agents but also on the durability of the rock mass concerned. This, in turn, is governed by the mineralogical composition, texture, porosity and strength of the rock on the one hand, and the incidence of discontinuities within the rock mass on the other hand. Hence, the response of a rock mass to weathering is directly related to its internal surface area and average pore size. Coarse-grained rocks generally weather more rapidly than fine-grained ones. The degree of interlocking between component minerals is also a particularly important textural factor, since the more strongly a rock is bonded together, the greater its resistance is to weathering. The closeness of the interlocking of grains governs the porosity of the rock. This, in turn, determines the amount of water it can hold, and hence, the more porous the rock is, the more susceptible it is to chemical attack. Also, the amount of water that a rock contains influences mechanical breakdown, especially in terms of frost action. Nonetheless, deep-weathered profiles usually have been developed over lengthy periods of time. The type and rate of weathering varies from one climatic regime to another. In humid regions, chemical and chemico-biological processes are generally much more significant than those of mechanical disintegration. On the one hand, the degree and rate of weathering in humid regions depends primarily on the temperature and amount of moisture available. An increase in temperature causes an increase in weathering. If the temperature is high, then weathering is extremely active; an increase of 10 ℃ in humid regions more than doubles the rate of chemical reaction. On the other hand, in dry air, chemical decay of rocks takes place very slowly.

Weathering leads to a decrease in density and strength, and to increasing deformability. An increase in the mass permeability frequently occurs during the initial stages of weathering due to the development of fractures, but if clay material is produced as minerals breakdown, then the permeability may be reduced. Widening of discontinuities in carbonate rock masses by dissolution leads to a progressive increase in permeability.

4.1.1 Mechanical Weathering

Mechanical or physical weathering is particularly effective in climatic regions that experience significant diurnal changes of temperature. This does not necessarily imply a large range of temperature, as the frost and thaw action can proceed where the range is limited.

Alternate freeze-thaw action causes cracks, fissures, joints and some pore spaces to be widened, as shown in Figure 4.1. As the process advances, angular rock debris is gradually broken from the parent body. Frost susceptibility depends on the expansion in volume that occurs when water moves into the ice phase, the degree of saturation of water in the pore system, the critical pore size, the amount of pore space, and the continuity of the pore system. In particular, the pore structure governs the degree of saturation and the magnitude of stresses that can be generated upon freezing (Bell, 1993). When water turns into ice, it increases in volume by up to 9%, thus giving rise to an increase in pressure within the pores it occupies. This action is further enhanced by the displacement of pore water away from the developing ice front. Once ice has formed, the ice pressures rapidly increase with temperature decreasing, so that at approximately −22 ℃, ice can exert a pressure of up to 200 MPa. Usually, coarse-grained rocks withstand freezing better than fine-grained ones. The critical pore size for freeze-thaw durability appears to be about 0.005 mm. In other words, rocks with larger mean pore diameters allow outward drainage and escape of fluid from the frontal advance of the ice line and, therefore, are less frost susceptible. Fine-grained rocks that have 5% sorbed water are often very susceptible to frost damage, whereas those containing less than 1% are very durable. Nonetheless, a rock may fail if it is completely saturated with pore water when it is frozen. Indeed, it appears that there is a critical moisture content, which tends to vary between 75% and 96% of the volume of the pores, above which porous rocks fail. The rapidity with which the critical moisture content is reached is governed by the initial degree of saturation.

The mechanical effects of weathering are well displayed in hot deserts where wide diurnal ranges of temperature cause rocks to expand and contract. Because rocks are poor conductors of heat, these effects are mainly localized in their outer layers where alternate expansion and contraction creates stresses that eventually rupture the rock. In this way, flakes of rock break away from the parent material, and this process is termed exfoliation. The effects of exfoliation are concentrated at the corners and edges of rocks, so that their outcrops gradually become rounded.

Figure 4.1 Freeze-thaw weathering

However, in hot semi-arid regions, exfoliation can take place on a large scale with large slabs becoming detached from the parent rock mass. Furthermore, minerals possess different coefficients of expansion, and differential expansion within a polymineralic rock fabric generates stresses at grain contacts and can lead to granular disintegration.

There are three ways whereby salts within a rock can cause its mechanical breakdown: by pressure of crystallization, by hydration pressure, and by differential thermal expansion (shown in Figure 4.2). Under certain conditions, some salts may crystallize or recrystallize to different hydrates that occupy a larger space (being less dense) and exert additional pressure, that is, hydration pressure. The crystallization pressure depends on the temperature and degree of supersaturation of the solution, whereas the hydration pressure depends on the ambient temperature

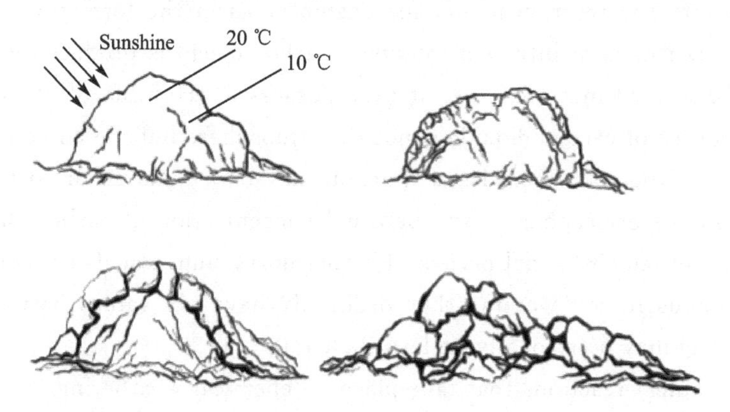

Figure 4.2 The weathering process caused by temperature difference

and relative humidity. Calculated crystallization pressures provide an indication of the potential pressures that may develop during crystallization in narrow closed channels. Crystallization of freely soluble salts such as sodium chloride, sodium sulphate or sodium hydroxide often leads to the crumbling of the surface of a rock such as limestone or sandstone. Salt action can give rise to honeycomb weathering in porous limestone or sandstone possessing calcareous cement.

4.1.2 Chemical and Biological Weathering

Chemical weathering leads to mineral alteration and the solution of rocks. Alteration is caused principally by oxidation, hydration, hydrolysis and carbonation, whereas solution is caused by acidified or alkalized waters. Chemical weathering also aids rock disintegration by weakening the rock fabric and by emphasizing any structural weaknesses, however slight, that it possesses. When decomposition occurs within a rock, the altered material frequently occupies a greater volume than that from which it was derived and, in the process, internal stresses are generated. If this expansion occurs in the outer layers of a rock, then it eventually causes them to peel off from the parent body.

In dry air, rocks decay very slowly. The presence of moisture hastens the rate of decay, firstly, because water is itself an effective agent of weathering and, secondly, because it holds in solution substances that react with the component minerals of the rock. The most important agents of these substances are free oxygen, carbon dioxide, organic acids and nitrogen acids. Free oxygen is an important agent in the decay of all rocks that contain oxidizable substances, iron and sulphur being especially suspect. The oxidation is speeded up by the presence of water; indeed, it may enter into the reaction itself, for example, as in the formation of hydrates. However, its role is mainly as a catalyst. Carbonic acid is produced when carbon dioxide is dissolved in water, and it may possess a pH value of about 5.7. The principal source of carbon dioxide is not the atmosphere but the air contained in the pore spaces in the soil where its proportion may be a hundred or so times greater than it is in the atmosphere. An abnormal concentration of carbon dioxide is released when organic material decays. Furthermore, humic acids are formed by the decay of humus in soil waters; they ordinarily have pH values between 4.5 and 5.0, but they may occasionally be less than 4.0.

The simplest reactions that take place on chemical weathering are the solution of soluble minerals and the addition of water to substances to form hydrates. Solution commonly involves ionization, for example, this takes place when gypsum and

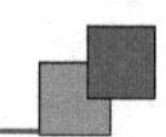

carbonate rocks are weathered. Hydration takes place among some substances. A common example is the reaction of gypsum turning into anhydrite:

$$CaSO_4 + 2H_2O = CaSO_4 \cdot 2H_2O$$

This reaction above produces an increase in volume of approximately 6% and, accordingly, causes the enclosing rocks to be wedged further apart. Iron oxides and hydrates are conspicuous products of weathering; usually the oxides are a shade of red and the hydrates yellow to dark brown.

Sulphur compounds are oxidized by weathering. Because of the hydrolysis of the dissolved metal ion, solutions forming from the oxidation of sulphides are acidic. For instance, when pyrite (FeS_2) is oxidized initially, ferrous sulphate and sulphuric acid are formed. Further oxidation leads to the formation of ferric sulphate. The formation of anhydrous ferrous sulphate can give rise to a volume increase of about 350%. Very insoluble ferric oxide or hydrated oxide is formed if highly acidic conditions are produced. Sulphuric acid may react with calcite to give gypsum that involves an expansion in volume of around 100%.

$$2FeS_2 + 7O_2 + 2H_2O = 2FeSO_4 + 2H_2SO_4$$

$$12FeSO_4 + 3O_2 + 6H_2O = 4Fe_2(SO_4)_3 + 4Fe(OH)_3$$

$$4Fe_2(SO_4)_3 + 6H_2O = 2Fe(OH)_3 + 3H_2SO_4$$

Perhaps the most familiar example of a rock prone to chemical attack is limestone. Limestones are mainly composed of calcium carbonate ($CaCO_3$). Aqueous dissolution of calcium carbonate introduces the carbonate ion into water, that is, $(CO_3)^{2-}$ combines with H^+ to form the stable bicarbonate H_2CO_3:

$$CaCO_3 + 2H_2O = Ca(OH)_2 + H_2CO_3$$

Weathering of the silicate minerals is primarily a process of hydrolysis. Much of the silica that is released by weathering forms silicic acid but, when liberated in large quantities, some of it may form colloidal or amorphous silica. Mafic silicates usually decay more rapidly than felsic silicates and, in the process; they release magnesium, iron and lesser amounts of calcium and alkalies. Olivine is particularly unstable, decomposing to form serpentine, which forms talc and carbonates on further weathering. Chlorite is the commonest alteration product of augite (the principal pyroxene) and of hornblende (the principal amphibole). The weathering process of orthoclase [$K(AlSi_3O_8)$] is as follow:

$$4K(AlSi_3O_8) + 6H_2O = 4KOH + 8SiO_2 + Ai_4(Si_4 + O_{10})(OH)_8$$

When subjected to chemical weathering, feldspars decompose to form clay minerals, which are, consequently, the most abundant residual products. The process is caused by the hydrolysing action of weakly carbonated water that leaches the bases out of the feldspars and produces clays in colloidal form. The alkalies are removed in solution as carbonates from orthoclase (K_2CO_3) and albite (Na_2CO_3), and as bicarbonate from anorthite [$Ca(HCO_3)_2$]. Some silica is hydrolysed to form silicic acid.

The colloidal clay eventually crystallizes as an aggregate of minute clay minerals. Deposits of kaolin are formed when percolating acidified waters decompose the feldspars contained in granitic rocks.

Plants and animals play an important role in the breakdown and decay of rocks, indeed their part in soil formation is of major significance. Tree roots penetrate cracks in rocks and gradually wedge the sides apart, whereas the adventitious root system of grasses breaks down small rock fragments to particles of soil size. Burrowing rodents also bring about mechanical disintegration of rocks. The action of bacteria and fungi is largely responsible for the decay of dead organic matter. Other bacteria are responsible, for example, for the reduction of iron or sulphur compounds.

4.1.3 Engineering Classification of Weathering

Weathering may change the physical and chemical properties of the rocks, destroy the integrity of the rocks, and weaken the capacity of the foundation and slope. The weathering degrees of the rock can be divided into six levels (shown in Table 4.1) for better application in engineering.

Table 4.1 Classification of weathering degrees of rocks

Weathering degrees	Characteristics	Weathering parameters	
		Wave velocity ratio K_v	Weathering coefficient K_f
Unweathered	Rock is fresh. Traces of weathering are very few	0.9~1.0	0.9~1.0
Weak weathering	Structure does not change, only the colour of cleavage plane has a slight discoloration. There is a small amount of weathering fractures	0.8~0.9	0.8~0.9
Medium weathering	Structural damages appear in some part. There are secondary minerals along joint plane. Weathering fractures are much more, rocks are cut into pieces. It is difficult to dig by pick	0.6~0.8	0.4~0.8

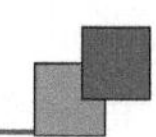

(continued)

Weathering degrees	Characteristics	Weathering parameters	
		Wave velocity ratio K_v	Weathering coefficient K_f
Intense weathering	Most of structures are damaged. Mineral compositions change significantly. Weathering fracture is developmental, rocks become crack. It is easy to dig by pick	0.4~0.6	<0.4
Weathering	Structures is failure and only have residual structural strength. It is easy to dig by pick	0.2~0.4	—
Residual soil	All of the structures are failure, rocks become soil by weathering	<0.2	—

Notes:

① Wave velocity ration, K_v, refers to the ratio of compressional wave velocity between weathered rocks and fresh rocks.

② Weathering coefficient, K_f, refers to the ratio of uniaxial compressive strength between weathered rocks and fresh rocks.

③ Type of granite rock can be divided according to the standard penetration test.

④ Weathering degrees of mudstone and half diagenetic don't need to be classified.

Weathering crust or weathered zone refers to the surface of the rock, which is divided into four weathering zones according to the weathering degrees of rocks (shown in Table 4.2).

Table 4.2 Basic characteristics of weathering rock zones

Weathering zones	Colour	Structure of rocks	Degree of crushing	Rock strength	Hammer sound
Completely weathered zone	Colour is changed completely	Structure is broken completely	It can be crushed into sand or soil by hand	Very low	Hoarse
Intensely weathered zone	Colour of many parts is changed	Most structures are destroyed	Loose, broken and poor integrity	Less than 1/3 of the original rock	Hoarse

(continued)

Weathering zones	Colour	Structure of rocks	Degree of crushing	Rock strength	Hammer sound
Weakly weathered zone	The colour of easily weathered minerals is changed partly	Structures are damaged partly	Weathering fractures are much more and integrity is poor	1/3 ～2/3 of the original rock	Sound is not clear
Slightly weathered zone	Only the part along the crack surface changes colour slightly	Structures have not been changed	There is a small amount of weathering fractures. It is difficult to distinguish with fresh rocks	Slightly lower than the fresh rock	Sound is ringing

4.1.4 Governance of Weathering

Weathering can reduce the strength of the rocks and the stability of the buildings. There are some commonly used methods to process weathering, such as dig method, isolation method of air, consolidation grouting method and drainage method.

① Dig method is a difficult and time consuming process, which is suitable for replacing the thin weathering layer. When the thickness of weathering zone is bigger, only upper part will be stripped.

② Isolation method of air can cover and protect stratum by using asphalt, cement and clay, etc.

③ Consolidation grouting method terms to strengthen the rock and reduce water permeability by pumping cement, asphalt, sodium silicate and clay into cracks or rocks.

④ Drainage method aims to weaken weathering velocity of rocks avoiding water.

4.2 River

4.2.1 River Erosion

The initial dominant action of master streams is vertical down-cutting that is

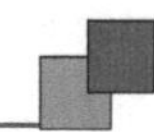

accomplished by the formation of potholes, which ultimately coalesce, and by the abrasive action of the load. Hence, in the early stages of river development, the cross profile of the valley is sharply V-shaped. As time passes, valley widening due to soil creep, slippage, rain-wash and gullying becomes progressively more important and, eventually, lateral corrasion replaces vertical erosion as the dominant process. A river possesses few tributaries in the early stages, but their numbers increase as the valley widens, thus affording a growing increment of rock waste to the master stream, thereby enhancing its corrasive power.

During valley widening, the stream erodes the valley sides by causingundermining and slumping to occur on the outer concave curves of meanders where steep cliffs or bluffs are formed. These are most marked on the upstream side of each spur. Deposition usually takes place on the convex side of a meander. Meanders migrate both laterally and downstream, and their amplitude is increased progressively. In this manner, spurs are eroded continuously, first becoming more asymmetrical until they are eventually truncated (shown in Figure 4.3).

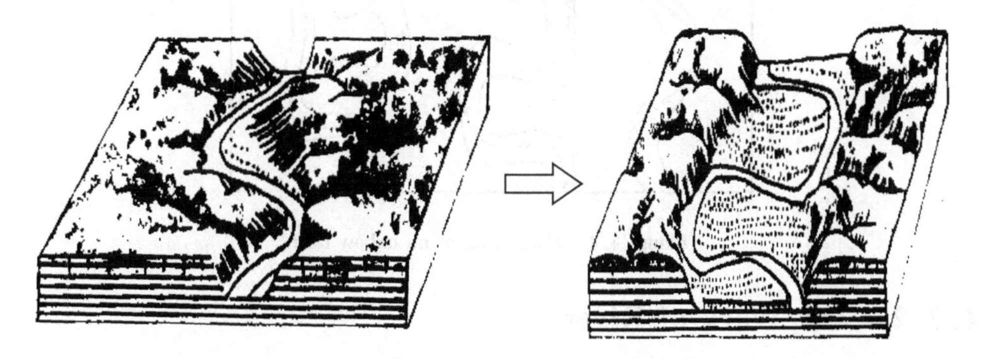

Figure 4.3 Widening of valley floor by lateral corrosion

The slow deposition that occurs on the convex side of a meander, as lateral migration proceeds, produces a gently sloping area of alluvial ground called the flood plain. The flood plain gradually grows wider as the river bluffs recede, until it is as broad as the amplitude of the meanders. From now onwards, the continuous migration of meanders slowly reduces the valley floor to an almost flat plain that slopes gently downstream and is bounded by shallow valley sides.

Throughout its length, a river channel has to adjust to several factors that change independently of the channel itself. These include the different rock types and structures across which it flows. The tributaries and inflow of water from underground sources affect the long profile of a river. Other factors that bring about adjustment of a river channel are flow resistance, which is a function of particle size, and the shape of transistor deposits such as bars, the method of load transport and the channel pattern including meanders and islands. Lastly, the river channel must also adjust itself to the river slope, width, depth and velocity.

Meanders, although not confined to, are characteristic of flood plains. The consolidated veneer of alluvium, spread over a flood plain, offers little resistance to continuous meander development. Hence, the loops become more and more accentuated. As time proceeds, the swelling loops approach one another. During the flood, the river may cut through the neck, separating two adjacent loops, thereby straightening its course. As it is much easier for the river to flow through this new course, the meander loop is silted off and abandoned as an oxbow lake (shown in Figure 4.4).

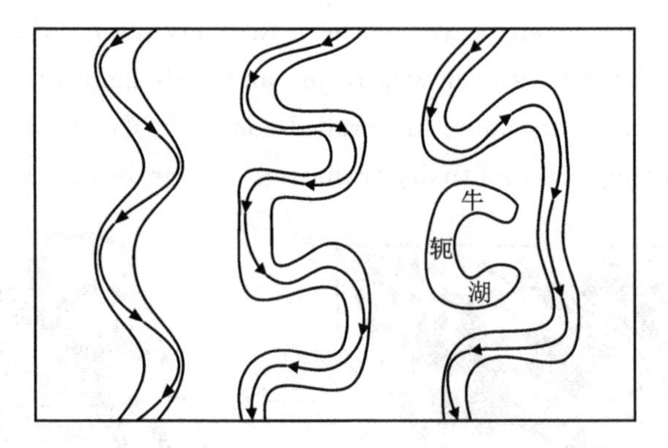

Figure 4.4 Formation of an oxbow lake

4.2.2 The Work of Rivers

The work done by a stream is a function of the energy it possesses. Stream energy is lost as a result of friction from turbulent mixing, and frictional losses are dependent on channel roughness and shape. Total energy is influenced mostly by velocity, which is a function of the stream gradient, volume and viscosity of water in flow and the characteristics of the channel cross section and bed.

The work undertaken by a river is threefold; it erodes soils and rocks and transports the products thereof, which it eventually deposits. Erosion occurs when the force provided by the river flow exceeds the resistance of the material over which it runs. The velocity needed to initiate movement, that is, the erosional velocity, is appreciably higher than that required to maintain the movement. Four types of fluvial erosion have been distinguished, namely, hydraulic action, attrition, corrasion and corrosion. Hydraulic action is the force of the water itself.

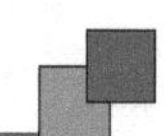

Attrition is the disintegration that occurs when two or more particles that are suspended in water collide. Corrasion is the abrasive action of the load carried by a river on its channel. Most of the erosion done by a river is attributable to corrasive action. Hence, a river carrying coarse, resistant, angular rock debris possesses a greater ability to erode than does one transporting fine particles in suspension. Corrosion is the solvent action of river water.

The load that a river carries is transported in four different ways as follows:

Firstly, there is traction, which is, rolling of the coarsest fragments along the river bed.

Secondly, smaller material, when caught in turbulent upward-moving eddies, proceeds downstream in a jumping motion referred to as saltation.

Thirdly, fine sand, silt and mud may be transported in suspension.

Lastly, soluble material is carried in solution.

Deposition occurs where turbulence is at a minimum or where the region of turbulence is near the surface of a river. For example, lateral accretion occurs, with deposition of a point bar, on the inside of a meander bend. The settling velocity for small grains in water is roughly proportional to the square of the grain diameter, whereas for larger particles, settling velocity is proportional to the square root of the grain diameter.

An alluvial flood plain is the most common depositional feature of a river. The alluvium is made up of many kinds of deposits, laid down both inside and outside the channel. Vertical accretion on a flood plain is accomplished by in-channel filling and the growth of overbank deposits during and immediately after floods. Gravel and coarse sands are moved mainly at flood stages and deposited in the deeper parts of a river.

As the river overtops its banks, its ability to transport material is lessened, so that coarser particles are deposited near the banks to form levees. Levees stand above the general level of the adjoining plain, so that the latter is usually poorly drained and marshy (shown in Figure 4.5). This is particularly the case when levees have formed across the confluences of minor tributaries, forcing them to wander over the flood plain until they find another entrance to the main river. Finer material is carried farther and laid down as backswamp deposits. At this point, a river sometimes aggrades its bed, eventually raising it above the level of the surrounding plain. Consequently, when levees are breached by flood water, hundreds of square kilometres may be inundated.

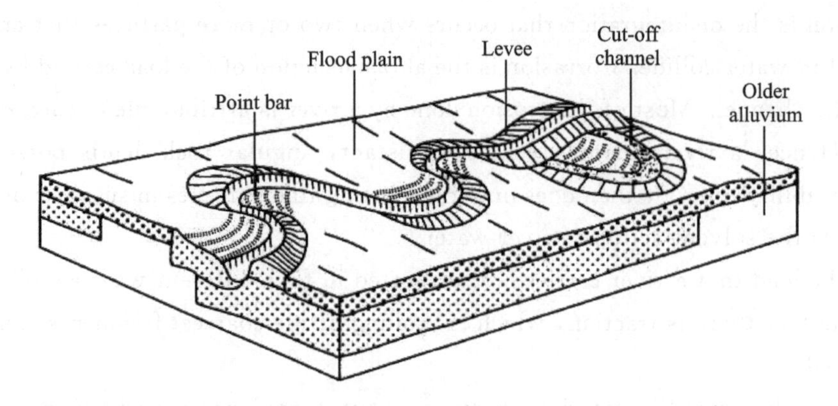

Figure 4.5 The main depositional features of a meandering channel

4.3 Groundwater

The principal source of groundwater is meteoric water, that is, precipitation. However, two other sources are occasionally of some consequence. These are juvenile water and connate water. The former is derived from magmatic sources, whereas the latter represents the water in which sediments are deposited. Connate water is trapped in the pore spaces of sedimentary rocks as they are formed and has not been expelled.

The retention of water in soil depends on the capillary force and the molecular attraction of the particles. As the pores in the soil become thoroughly wetted, the capillary force declines, so that gravity becomes more effective. In this way, downward percolation can continue after infiltration has ceased but the capillarity increases in importance as the soil dries. No further percolation occurs after the capillary and gravity forces are balanced. Thus, water percolates into the zone of saturation when the retention capacity is satisfied.

4.3.1 Vadose Water

The water in the zone of aeration is termed as vadose water (shown in Figure 4.6). A perched water table is one that forms above a discontinuous impermeable layer such as a lens of clay in a formation of sand, the clay impounding a water mound. Vadose water affects some construction projects, such as foundation pit, tunnel and road.

The supply of vadose water is mainly from rain, so that it only can provide a little water to use Water pollution should be paid attention to.

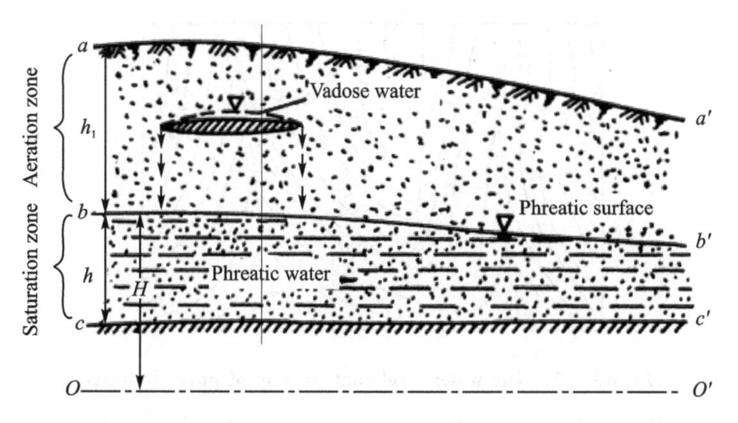

aa'—Earth surface; *bb'*—Phreatic surface; *cc'*—Aquitards surface; *OO'*—datum plane

Figure 4.6 Vadose water and phreatic water

4.3.2 Phreatic Water

If pores insaturation zone are filled with water, the water is generally termed as phreatic water. The upper surface of this zone is therefore known as the phreatic surface. The elevation of the phreatic surface is termed as water table. The vertical distance between phreatic surface and the Earth surface is termed as buried depth. The vertical dimension between phreatic surface and bottom board of aquiclude water is termed as the thickness of the phreatic aquifer.

Phreatic water can flow from high water level to low water level under the action of gravity and has free water surface. The physiognomy of phreatic water is controlled by topography, which is basically identical with topography, and gentler than terrain. Phreatic water is greatly influenced by climate, which has obvious seasonal variation characteristics, and is easy to be affected by ground pollution.

The water level contour map of phreatic water refers to the connection diagram of the equal elevation of phreatic surface, which needs to indicate the date of the determination of water level. In general, water level contour maps of lowest water level and highest water level are needed. The water level contour maps can solve the following questions.

(1) Judging the Flow Direction of Phreatic Water

In water level contour maps of phreatic water, the direction perpendicular to the contour is termed as the direction of phreatic water, and the arrow points to the low water level. For example, in Figure 4.7, the direction of phreatic water is from point E to point F.

(2) Calculating Hydraulic Gradient

The hydraulic gradient refers to the ratio of water head of two points and hori-

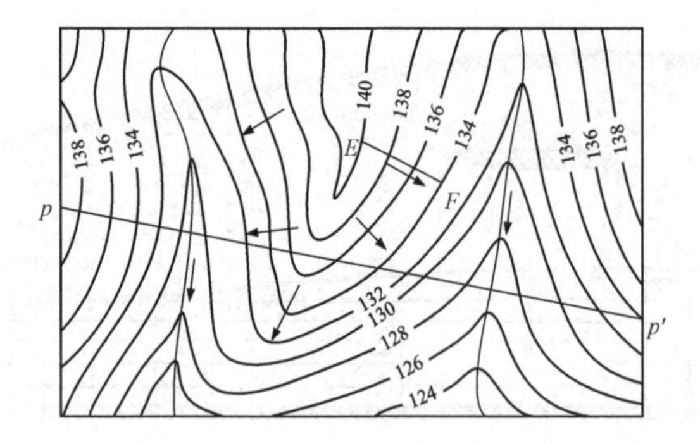

Figure 4.7 The water level contour map of phreatic water

zontal distance. For example, in Figure 4.7, the hydraulic gradient of line EF is

$$I_{EF}=\frac{140-134}{1\ 000}=0.006$$

(3) Judging Water Supply Relationship

If the flow direction of phreatic water is from river to outside, the river supplies phreatic water (shown in Figure 4.8(a)). If phreatic water runs into the river, the phreatic water supplies river (shown in Figure 4.8(b)).

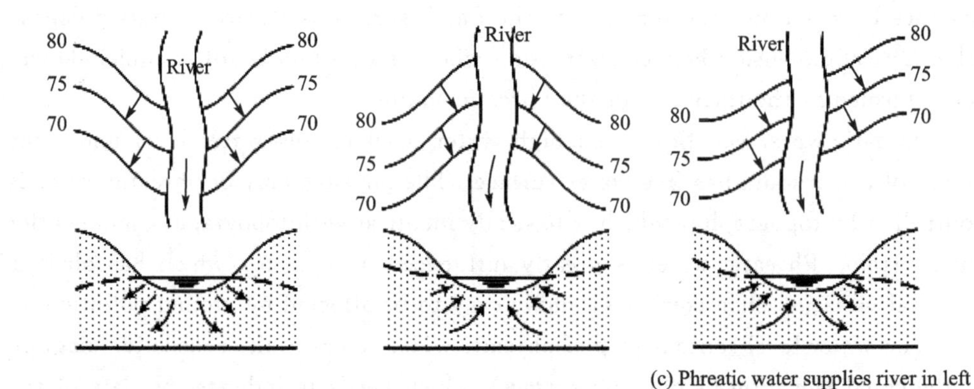

(a) River supplies phreatic water (b) Phreatic water supplies river (c) Phreatic water supplies river in left side of the river and river supplies phreatic water in right side

Figure 4.8 The supply relationship between river and phreatic water

(4) Calculating the Buried Depth of Phreatic Water

The difference value between the elevation of some points in topographic contour and the elevation of the water level contour map of phreatic water is termed as the buried depth of phreatic water.

(5) Judging the Position of Spring or Swamp

If the phreatic water appears in the position where contour of water table of phreatic water and topographic contour have equal elevation, it is spring or swamp.

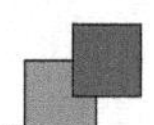

4.3.3 Aquifers, Aquicludes and Aquitards

An aquifer is the term given to a rock or soil mass that not only contains water but from which water can be abstracted readily in significant quantities. The ability of an aquifer to transmit water is governed by its permeability. Indeed, the permeability of an aquifer usually is in excess of 10^{-5} m/s.

By contrast, a formation with a permeability of less than 10^{-9} m/s is one that, in engineering terms, is regarded as impermeable and referred to as an aquiclude. For example, clays and shales are aquicludes. Even when such rocks are saturated, they tend to impede the flow of water through stratal sequences.

An aquitard is a formation that transmits water at a very slow rate but that, over a large area of contact, may permit the passage of large amounts of water between adjacent aquifers that it separates. Sandy clays provide an example.

4.4 Land Subsidence

Groundwater is an important part of the geological environment and the most active one of external geological factors. In many cases, the change of geological environment is usually caused by the change of groundwater. There exist many factors for the change of groundwater, which can be summarized as natural factors and human factors. Natural factors mainly refer to the climate factors, such as the change of groundwater caused by precipitation, which is typically a large range event and could be predicted. The human factors for the change of the groundwater is various, often with a contingency, which is hard to be predicted and does significant harm to the engineering practice.

Land subsidence is also called surface subsidence, refers to the ground settlement of a certain range of horizontal plane surface. The settlement is one kind of non-compensable, permanent environmental and resource loss.

4.4.1 Reasons of Land Subsidence

1. Geological Factors

① Under the action of gravity, loose stratum become hard or half-hard rocks and land subsidence starts because of the decrease of bed thickness.

② The tectonic movement including horizontal motion and lifting movement can lead to the ground sinking and then land subsidence.

③ The earthquake causes land subsidence. Seismic subsidence refers to the settlement of engineering structures or ground caused by the increase of soil density, extension of plastic zone or strength decrease under the strong earthquake.

2. Human Factors

① The excessive exploitation of groundwater is the main cause of land subsidence.

② The excessive exploitation of underground resources such as oil, gas and solid mineral is also the cause of land subsidence.

③ The load of the surface becomes large with the increase of high buildings, railways, bridges and other transportation facilities, which can speed up the settlement of ground to some extent.

4.4.2 Mechanism of Land Subsidence

The generation of land subsidence can be summarized as follow: stress in the overlying strata is borne by the soil particle skeleton and water together. The part of the stress soil particle skeleton bearing is called the porosity water pressure. When groundwater is extracted from the aquifer, the pressure level drops so as the porosity water pressure. To achieve the balance, the effective stress increases. The increment in the effective stress has an impact on both the aquifer and aquicludes, namely, the consolidation compression in between, eventually resulting in the land subsidence. Deformation characteristics of the aquifer and water-resisting layer are different. Firstly, the compression ratio of cohesive soil is 1～2 orders of magnitude larger than the sandy soil, so the consolidation compression of water-resisting layer is the main cause of land subsidence. Secondly, the deformation due to water-release compression of sandy soils is elastic, therefore, the corresponding land subsidence caused is temporary. After the aquifer acquires extra water, pore water pressure level rises. Meanwhile, the effective stress drops, leading to the rebound of aquifer. Thirdly, the deformation due to water-release compression of cohesive soil is plastic, and the aquifer deformation is permanent which cannot bounce back after the water supplement.

4.4.3 Harm of Land Subsidence

Land subsidence causes the destruction of the surface buildings and underground facilities. According to statistics, China's annual economic loss caused by

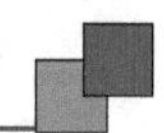

land subsidence is more than hundreds of millions RMB. The main damage are listed as follows:

① Land subsidence leads to ground fissure, which may imperil urban and rural areas.

② Land subsidence jeopardizes the city pipe network. It causes the bending deformation of urban water supply pipe, gas pipe, wire and optical cable, which directly affect the public life and industrial production.

③ Land subsidence causes the railway roadbed to subside unevenly, threatening the safety of railway.

④ Land subsidence causes the riverbed to subside, reducing flood storage capacity of the river.

⑤ Land subsidence cause s sea water encroachment, reducing the value of land.

⑥ The ground elevation is an important benchmark for city surveying and mapping, city planning as well as construction. Land subsidence invalids the benchmark of the ground, ground elevation information, and the sign of observations and measurements.

4.4.4 Measures of Controlling Land Subsidence

In order to control land subsidence, areas of excessive exploitation of groundwater should be primarily monitored and governed. Main measures are listed as follows.

(1) Water Saving Measures

For example, in America, people use advanced transportation method of groundwater to make long-term plan on groundwater usage, in order to save water.

(2) Artificial Recharge of Groundwater

Raise the underground water level by supplementing the groundwater. However, the pollution of groundwater needs to be paid attention to in this process.

(3) Reinforce Embankment

Reinforce the coast of coastal cities, such as heightening and strengthening flood wall, sea embankment, floodgates, damp-proof, and so on.

(4) Strengthen Monitoring and Investigation of Land Subsidence

Investigate the current status of land subsidence, researching the subsidence mechanism, forecasting speed, scope and range of land subsidence and establishing early warning mechanism.

Thinking Questions

1. Please explain the geological process of the river.
2. Please explain aquifer and aquiclude.
3. Please explain the groundwater types according to the burial conditions.
4. Please list the uses of the water level contour map of phreatic water.
5. Please explain the mechanism, formation causes and harms of land subsidence.

Chapter 5

Common Geological Disasters

Geological disaster, refers to the geological events, caused by natural geological processes and human activities, which results in the deterioration of the geological environment quality, directly or indirectly harm human security, and causes loss to the society and economic development, such as land subsidence, landslide, debris flow, earthquake and Karst.

5.1 Landslide

Landslide refers to the phenomenon that soil of slope and rock loses the original steady state under the gravity, and slides down along the slope within a certain sliding surface. It is a common geological disaster in mountain areas and is one of the most destructive adverse geologic phenomena. Firstly, rock and earth mass sliding retains the original rock integrity except for fewer collapses and fracture features in edge areas and some local areas of the landslide. Secondly, the movement of rock mass on the slope is sliding, not dumping or rolling, therefore, the lower limb of landslide is often slip surface or the location of the sliding zone. In addition, some landslides move really quickly, and roll-over phenomenon may occur on the surface of landslide body. This phenomenon is called collapse landslide.

On some steep or extremely steep slopes, some large or giant pieces of rock burst or slide suddenly, jumping along the hillside fiercely. Rock masses bump into each other and then break into pieces, accumulating in the slope toe finally. The whole process is known as collapse, and accumulation mass in the slope toe is called

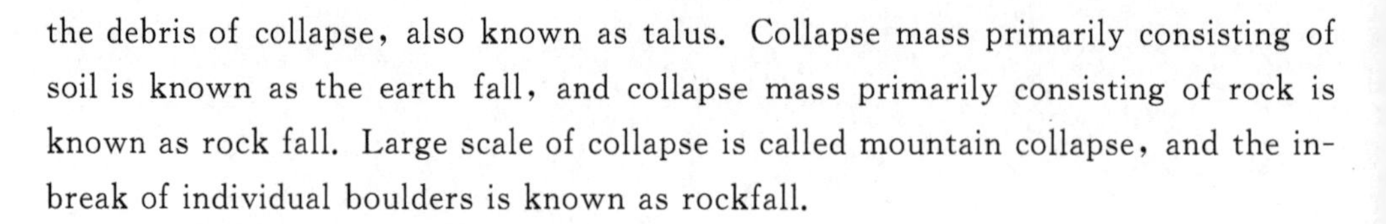

the debris of collapse, also known as talus. Collapse mass primarily consisting of soil is known as the earth fall, and collapse mass primarily consisting of rock is known as rock fall. Large scale of collapse is called mountain collapse, and the in-break of individual boulders is known as rockfall.

5.1.1 Elements of Landslide

A typical fully developed landslide has the basic elements in below (shown in Figure 5.1).

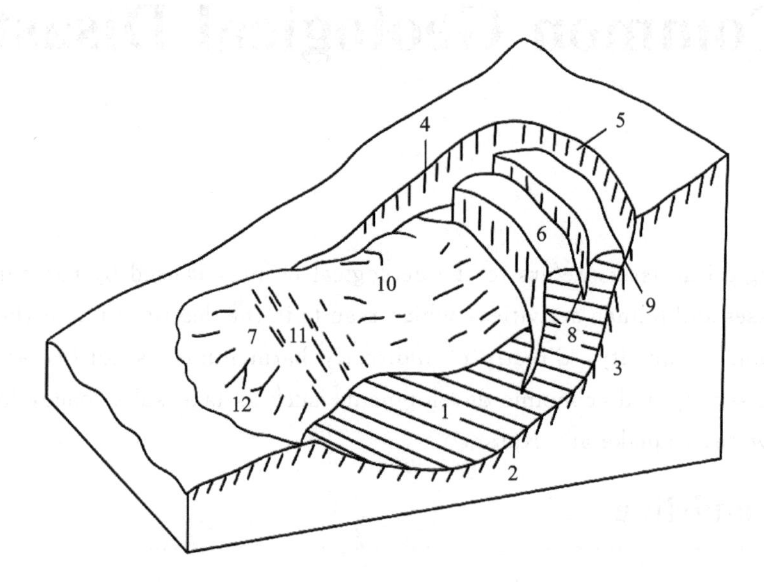

1—Sliding body; 2—Slide surface; 3—Sliding bed; 4—Slide boundary;
5—Sliding cliff; 6—Sliding terrace; 7—Sliding tongue; 8—Tension crack;
9—Main crack; 10—Shear crack; 11—Convex crack; 12—Fan-shaped crack

Figure 5.1 Schematic diagrams of landslide elements

(1) Sliding Body

Sliding body refers to rock mass sliding along the slide surface in slope. The volume of landslide mass varies, and large landslide mass can amount to tens of millions of cubic meters.

(2) Sliding Surface

Sliding surface, also known as sliding bed surface or sliding surface refers to continuous failure surface connecting with free face, which is the interface between sliding body and sliding bed. Many slide surfaces can form sliding zone.

(3) Sliding Bed

Sliding bed refers to the still rock mass underlying the slide surface. This part

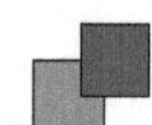

of rock mass keeps the original structure completely.

(4) Sliding Boundary

Sliding boundary refers to the line of demarcation between the sliding body and surrounding rock mass.

(5) Sliding Cliff

Sliding cliff refers to the upper limb of sliding surface. Sliding cliff looks like a round-backed armchair on the plane.

(6) Sliding Terrace

Sliding terrace refers to ladder-like ground formed because different parts of sliding body slide down with different speeds and amplitudes.

(7) Landslide Drumlin

Landslide drumlin refers to the uplift of hill formed when the moving sliding body encounters barriers.

(8) Sliding Tongue

Sliding tongue refers to the front part of the landslide, which stretches forth just like a tongue.

(9) Landslide Spindle

Landslide spindle, also known as the main sliding line, is the fastest vertical line with which the landslide moves along and is on behalf of the sliding direction of the landslide. The slide trace can be a straight line or a fold line.

(10) Landslide Cracks

During the landslide movement, cracks are generated due to different moving speeds of different parts of the landslide in the landslide body and surface. These kinds of cracks are called landslide cracks. According to different stress state, the landslide cracks can be divided into four kinds in below:

① Tension crack. When the slope is about to slide, some gaping arc-shaped cracks is generated in the rear of the sliding body because of tension force. Tension cracks coinciding with back scarp are called main cracks. Those main cracks appearing on the slope is the sign of landslide.

② Shear crack. When both sides of the sliding body start the relative displacement towards the adjacent steady rock mass or the former moves faster than the latter, there will be shearing actions which lead to cracks roughly parallel to the sliding direction. Either side of these cracks is accompanied by pennant, parallel, less-level cracks.

③ Convex crack. During the slide, the lower part of the landslide will bulge upward and crack if there is a resistance or the upper part slides too fast. These cracks are normally open and the lineup direction is perpendicular to the sliding direction.

④ Fan-shaped crack. During the downward sliding, sliding tongue spreads to both sides and form radial open cracks, which are fan-shaped cracks or front-edge radial cracks.

5.1.2 Classification of Landslide

In order to recognize and control landslide, the classification of landslides is needed. Because natural geological conditions and effects are complex and the purpose of all kinds of engineering classification and requirements is not the same, we can classify the landslide from different perspectives. Landslides in China can be classified as shown in Table 5.1.

Table 5.1 Types and characteristics of landslide

Standard of classification	Name	Characteristics
Lithology type of slope	Cohesive soil landslide	Sliding surface is roughly arc-shaped
	Loess landslides	
	Cracking rock landslide	
	Rock landslide	
	Landfill soil landslide	
	Deposit landslide	
Relation between sliding surface and bed plane	Homogeneous landslide	It occurs in soft stratum such as mudstone, shale and marl, and has a smooth sliding surface
	Incision landslide	Landslide that sliding surface is tangent with bed plane and incision landslide in rock mass where the hard stratum alternates with the soft stratum
	Bedding slide	Sliding along the bed plane or plane of fracture or the interface of diluvial layer and bed rock or plane of unconformity in bed rock
Genetic type of main slide surface	Accumulated layer landslide	
	Bed plane landslide	
	Structural surface landslide	
	Same-surface landslide	

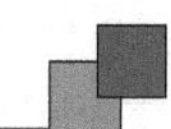

(continued)

Standard of classification	Name	Characteristics
Sliding velocity	Creep landslide	It cannot be observed by naked eyes but can be seen by apparatus
	Low-speed landslide	It slips a few centimeters to tens of centimeters every day and is visible to naked eyes
	Medium-speed landslide	It slides tens of centimeters per hour
	High-speed landslide	It slides several meters to tens of meters per second
Landslide depth	Shallow landslide	<6 m
	Medium-depth landslide	$6\sim20$ m
	Deep landslide	$20\sim50$ m
	Ultra-deep landslide	>50 m
Landslide scale	Small-scale landslide	$<10^5$ m^3
	Medium-scale landslide	$10^5\sim10^6$ m^3
	Large-scale landslide	$10^6\sim10^7$ m^3
	Giant landslide	$>10^7$ m^3
Forming age	New landslide	Landslide formed by excavating mountain
	Ancient landslide	Stable period of sliding body more than ten years
	Old landslide	Stable period of sliding body is $2\sim3$ years
	Developing landslide	
Mechanical condition	Retrogressive landslide	Lower part slides first, inducing the sliding of the upper part
	Push-type landslide	The upper part slides first and then, it extrude lower part to slide
	Laterally moving landslide	Local slide induces global sliding of entire body
	Hybrid landslide	Initial sliding caused by the collaborative move of upper and lower parts
Material composition	Earth slide	
	Rock landslide	
Sliding form	Rotation type landslide	
	Translation type landslide	

5.1.3 Factors of Landslide

Causes of slope instability are called landslide factors. They are introduced in below.

1. Appearance of a Slope

The existence of the slope can make slip surface expose in foreslope, which is the precondition for landslide. Generally speaking, slopes of river, lake (reservoir), sea, ditch and front open hillsides, railways, highways and engineering structure are geomorphologic positions where landslides are prone to happen. At the same time, factors like height, tilt, as well as the shape of slope can change the internal force state of slope, which results in the slope instability. The steeper and higher the slope becomes, the more easily the landslides will happen. When the upper part of slope protrudes and lower part concaves, and no anti-sliding terrain in slope toe exists, the landslide is more likely to occur.

2. Lithology of a Slope

Rock and earth mass is the material basis for the landslide, and various types of rock and soil are likely to constitute the sliding body. Generally, the higher the mechanical strength of rock and earth mass is, and the more complete the rock and earth mass is, then the less the possibility of landslide will be. Conversely, if the rock and earth mass structure is loose, the shear strength and weathering resistance of materials is low, such as loose covering layer, loess and red clay, shale, mudstone, etc. The landslides are prone to happen. Geotechnical properties of landslide slip surface directly affect the sliding speed. The lower the mechanical strength of the landslide is, the higher the land sliding speed will be.

3. Geological Structure of a Slope

When rock and earth mass is cut into discrete state by all kinds of structural surface, the slope may start to slip. The further development of the geological structure, the greater formation scale of the landslide. At the same time, the structural surface provides water with infiltration channel. Therefore the slope developed from joint, level, fault especially parallel and vertical to the structural plane of slope having a steep dip angle and slope of the structure of the flat surface is prone to become a landslide.

4. Water

Water can soften rocks, reduce the rock strength and speed up the weathering.

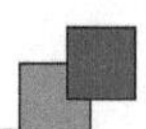

In addition, surface water can also flush and erode slope toe. Dynamic water pressure and pore water pressure produced by groundwater can subsurface erode rock and earth mass, and apply buoyancy on the permeable strata. A lot of landslides have characteristics called "big rain leading to large sliding, little rain leading to small sliding, no rain leading to no slip".

5. Earthquake

Earthquake can trigger landslides, and this phenomenon is very common in the mountain area. Firstly, earthquake breaks the structure of slope rock mass and makes the powder sand layer liquefy, thus reduce the shear strength of rock and soil body. Secondly, earthquake wave transmits in rock and soil, and applies the inertia force on rock mass, which increases the landslide slip force and triggers landslides.

6. Human Factors

(1) Excavation on Slope Toe

Railways, roads and other projects often slide during the construction because the lower part of slope lacks support.

(2) Water Storage and Drainage

Overflow and leakage of canals and ponds, water use and wastewater discharge of industrial production, agricultural irrigation and so on are easy to make water seeping into the slope body, increase porosity water pressure, soften rock mass, and increase the density of slope body, which eventually induces a landslide.

(3) Ramping Loading

Landfill rock and soil mass on upper part of a slope will increase the load and undermine the stability of the slope.

5.1.4 Treatment of Landslide

Landslide prevention and control have to implement the principle "early detection and prevention, full understanding and comprehensive management; total mitigation and no future trouble" and combine the factors of the slope instability and the internal and external conditions of landslide formation. In China there are a lot of engineering measures to prevent and control landslide which can be listed in three types as follows:

① Eliminate or reduce water hazards;

② Improve the mechanical properties of rock mass;

③ Improve the properties of sliding surface.

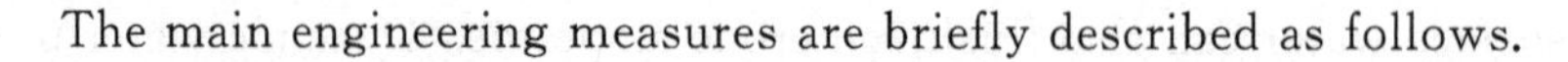

The main engineering measures are briefly described as follows.

1. Eliminate or Diminish Danger Caused by Water

(1) Elimination of Surface Water

Elimination of surface water is an indispensable auxiliary measure to control landslide, and should be the primary and long-term measure. The purpose of this treatment is to intercept or side lead surface water near the landslide zone; avoid surface water flow into the landslide area and eliminate rainwater and spring water in landslide area as soon as possible; prevent rainwater and spring water to penetrate into the landslide body. Main engineering measures include setting intercepting ditch out of the landslide body, and drains, spring project on landslide surface as well as landslide area greening engineering.

(2) Elimination of Groundwater

The better treatment for groundwater is to drain not to block. The main engineering measures can be listed as follows:

① Blind ditch for water interception is used to intercept and lead groundwater on the periphery of landslide area.

② Blind ditch for support is used for both drainage and supporting.

③ Tilted multiple holes, nearly horizontal drilling holes, lead groundwater out of underground soil body.

In addition, there are blind holes, permeability tube, vertical drilling, which are used to lead the groundwater out of the landslide body too.

(3) Prevention of the River Slope Toe Erosion

To prevent water to flush the landslide slope toe, the main engineering measures are: construct spur dike on the upstream landslide seriously eroded to lead mainstream to the other side; riprap, pave stone cage, build reinforced concrete pipe on landslide front to prevent slope toe from the river erosion.

2. Improve the Mechanical Condition of Rock and Soil Mass

(1) Cut Slope to Lose Weight

This method is often used in governing sliding body which is in a state of "top-heavy and bottom-light" and lack of reliable anti-sliding section in the front. This method can lower the center of gravity of sliding body and improve the shape, thus improve the stability of the landslide mass, as shown in Figure 5.2. The treatment of cutting slope should manage to reduce the height of the unstable rock mass, avoiding to cut resistance part of the rock mass.

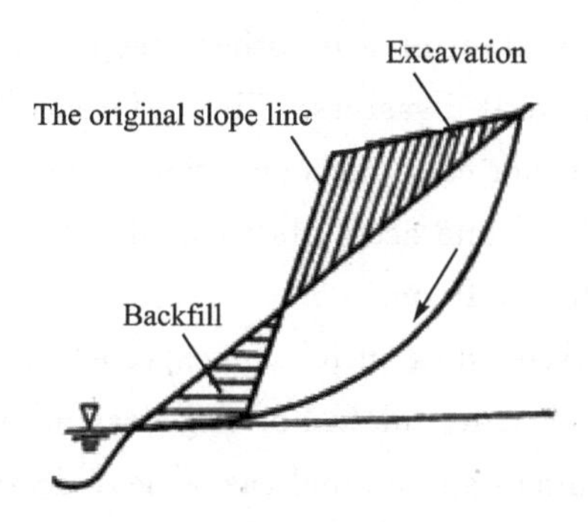

Figure 5.2 Schematic diagram for cutting slope

(2) Construct Retaining Elements

Because landslides resulting from losing support are typically steep and slide faster, we can build retaining elements to increase gravity balance condition of landslides and restore stability of sliding body quickly. Commonly artificial methods for reinforcing slope are as follows:

① Build retaining wall, curtain wall, etc. and support instable rock mass.

② Use reinforced concrete slide-resistant pile or reinforced pile as slide-resistant support elements.

③ Pre-stressed bolts or cables, which are suitable for reinforcing the rock slope containing inherent fissures or weak structural interface.

④ Consolidation grouting or electrochemical reinforcement method to strengthen the slope rock mass or soil mass.

⑤ Slope SNS flexible protection technology.

⑥ Set groove seam. In order to prevent the crack, sewing, and further development of the hole, use flaky to fill the empty crack and sewing of slope and cement mortar to groove seam, etc.

3. Improve the Properties of Sliding Surface

Directly stabilize the landslide by roasting, electro-osmotic drainage, grouting and chemical reinforcement, etc.

Due to the complexity of causes and influential factors of landslide, we often need to use several methods simultaneously and take comprehensive treatments to achieve the purpose of stabilizing the slope.

5.2 Debris Flow

Debris flow refers to special torrent with a large amount of mud, sand and stones occurring in the mountain area or other steep terrain because of the heavy rain, blizzard, or other natural disasters. The difference between debris flow and general flood is that the former contains a sufficient amount of mud, sand and other solid debris, and the mud volume accounts for at least 15% and up to 80%. It is more destructive than ordinary flood.

Debris flows have characteristics such as unexpected occurrence, fast flow rate, large flow, large capacity of material and strong destructive power. It can destroy cities and towns, cause human and animal casualties, destroy forest and farmland, and silt up the river, etc.

There are more than 50 countries threatened by the potential debris flow in the world, such as Colombia, Peru, Switzerland, Japan and China. China has more than 10 000 debris flow gullies and more than 70 counties are threatened by the potential debris flows. Rainwater debris flow are mostly distributed in Tibet, Sichuan, Yunnan and Gansu, while snow and ice debris flow are mainly distributed in the Tibetan plateau.

5.2.1 Formation Conditions of Debris Flow

The formation of debris flow is closely related with natural conditions of locating area and human economic activities.

1. Geological Condition

The development area of the debris flow has a complex geological structure and weak lithology, which is also subjected to intensive weathering action as well as frequent earthquake activities. Therefore, there are a lot of broken rocks mass in this area which provides a rich solid material for debris flow.

2. Topographic Condition

Areas prone to debris flow have deep grooves and high mountains, having a steep and longitudinal slope. Gully bed has a characteristic of basin shape that is good for collection. On the landscape, the area of general mud-rock flow formation can be divided into three parts: forming area, circulation area and accumulation area, as shown in Figure 5.3.

(1) Forming Area

Forming area is generally located in the upstream of rivers, and is surrounded by mountains on three sides and the forth side is gourd ladle or funnel-shaped. Surrounding mountains are steep, having a slope angel of 30° ~ 60°. Slope body is often bare, and vegetation is undergrowth. Slope is often gully-cutting, causing the

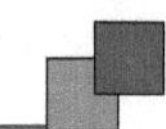

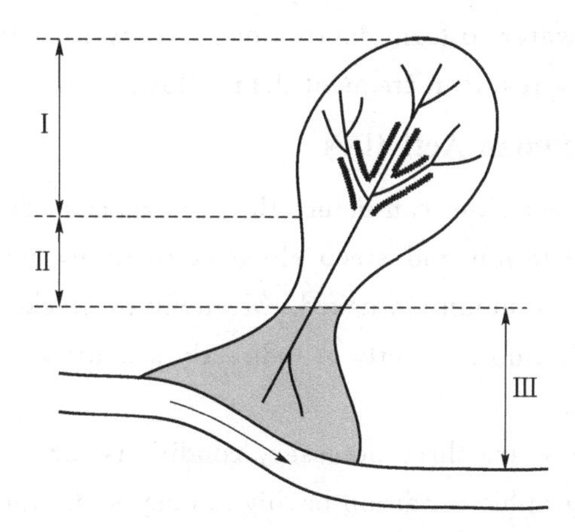

Ⅰ—Forming area; Ⅱ—Circulation area; Ⅲ—Accumulation area

Figure 5.3 Schematic diagrams for debris flow partition

collapse and development of landslides. The terrain is in favor of the concentration of water and debris.

(2) Circulation Area

Circulation area is where debris flow carries through and is narrow and deep valley or gully with steep valley walls, having big slope and many scarps. Mudslides flowing into this area gain a strong ability to scour, and wash down rock from the gully bed and carry them away. On the one hand, when longitudinal slope in circulation area is steep and straight, unimpeded debris flow can flow without obstacles, causing great harm. On the other hand, the energy will be weaken by easy jam or rechanneling.

(3) Accumulation Area

Accumulation area is the stacking zone for debris flow materials, generally located in the mountain or edge of intermountain basin. Because of suddenly open flat terrain, the kinetic energy of debris flow decreases sharply. Eventually debris flow stops and becomes fan-shaped deposit, conical-shaped deposit and band-shaped deposit namely the diluvial fan. When the diluvial fan is stable and not expanding, the destructive force of the landslide starts to decrease, and finally disappears.

3. Hydrology and Weather Condition

Hydrology and weather conditions of debris flow refer to a lot of collected water brought to the valley areas by heavy rains. Water is both a part of the debris flow and the debris flow transport medium. Loose solid material after filling a large amount of water becomes saturated or oversaturated and has a damaged structure. Friction resistance is reduced and liquidity increases. Therefore the loose solid ma-

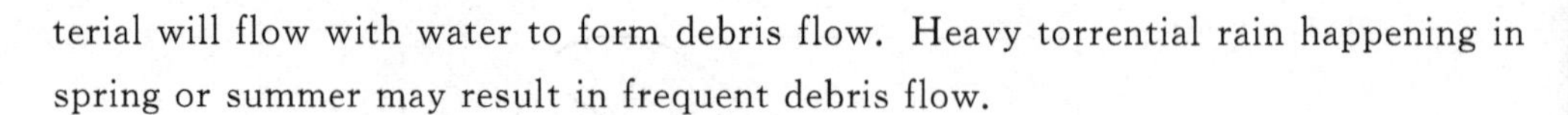

terial will flow with water to form debris flow. Heavy torrential rain happening in spring or summer may result in frequent debris flow.

4. Influence of Human Activities

Improper human activities can induce the occurrence of debris flow and intensify it, such as deforestation and steep slope cultivation, which severely damage vegetation and make the mountain naked. Meanwhile, stacked waste slags generated by mining, road-building directly or indirectly provide the materials for the formation of debris flow.

In summary, there are three necessary conditions for the formation of debris flow: favorable topographic condition having a steep slope for collecting water and solid materials, rich loose solid materials stacked in upstream, a great amount of flow water in a short and sudden period.

5.2.2 Classification of Debris Flow

1. By Material Composition

① Debris flow—consists of a large amount of cohesive soils and sand/stones of different size.

② Mud flow—consists of mainly cohesive soils and a small amount of sand and stones. This kind of debris flow has big viscosity and is like thick mud.

③ Water debris flow—consists of water and various sizes of sand and stones.

2. By Watershed Morphology

① Standard debris flow—typical debris flow can obviously be divided into formation area, circulation area and accumulation area.

② Valley-type debris flow—basin looks like a long and narrow strip. Formation area is usually the gully in upstream of river. Circulation area and accumulation area cannot be separated obviously.

③ Slope debris flow—basin looks like a hopper and no obvious circulation area exists. Formation area and accumulation area are directly connected.

3. By State of Matter

① Viscous debris flow—debris flow or mud flow, with a large amount of cohesive soil, has characteristics of high viscosity. 40% ~ 60%, even 80% of the flow, is solid material. It has a high solidity and stone particles are in suspended state. This kind of debris flow happens suddenly, lasts shortly and has a strong destructive force.

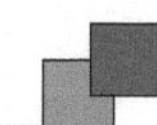

② Diluted debris flow—water is the main composition and little cohesive soil exists. 10% ~ 40% of the flow is solid material. This kind of debris flow has a great dispersibility. Water is the transport medium. Stones move forward in scroll or saltation way, having a strong incised effect.

4. By Amount of Solid Matter

① Giant debris flow—total solid matter in a debris flow is greater than 50×10^4 m^3.

② Large landslides—total solid matter in a debris flow is $10\times10^4\sim50\times10^4$ m^3.

③ Medium mud—total solid matter in a debris flow is $1\times10^4\sim10\times10^4$ m^3.

④ Small debris flow—total solid matter in a debris flow is less than 1×10^4 m^3.

In addition to the above classified methods, debris flow can also classify into river-type and rainfall-type debris flow by the formation cause. Debris flow can be also classified into debris flow in development period, in flush period and in development period.

5.2.3 Prevention and Control of Debris Flow

(1) Crossing Project

Build bridges above debris flow gully to allow debris flow discharges below. This is a commonly used measure in railway and highway transportation departments.

(2) Traversing Project

Build tunnel, open-cut tube or launder beneath the debris flow. Debris flow discharges above the facilities. This is a major project form for railway and highway through the debris flow area.

(3) Protecting Project

Build protective structures for bridges, tubes, substructure and so on to resist or eliminate harms such as scour, impact, lateral erosion caused by debris flow to main buildings. This kind of project mainly includes slope protection, barricade, spur dike and so on.

(4) Drainage Project

Build diversion dike, torrent gutter, beam barrier and so on to improve streaming potential of debris flow and increase discharge capacity of structures such as bridges, and to ensure the debris flow discharge smoothly according to the design intentions.

(5) Blocking Project

Build block slag dam, reservoir silt field, retaining engineering, flood control engineering to control the solid material, heavy rain and flood runoff of debris flow and decrease flow, drainage quantity and energy of debris flow. This project can reduce damages such as scour, impact and bury caused by debris flow.

To prevent and control of debris flow, a combination of various measures of treatment is frequently used.

5.3 Karst

Karst is a series of geological phenomena generated by erosion of soluble rock by surface water or groundwater. Karst is originally a name of limestone plateau of Istria peninsula in the northwest part of Yugoslavia, where has a typical Karst landform. Karst is closely related to the project construction. It can cause water leakage, tunnel water gushing, and subgrade damage and so on. A lot of attention should be paid to the Karst.

5.3.1 Form of Karst

1. Surface Configuration

(1) Karren, Clint and Stone Forest

When surface water flows along the surface of rock, many grooves are formed due to corrosion and erosion. These grooves are called Karren. Prominent part in Karrens is called a Clint. Stone forest is a kind of tall Clints with a height up to 20 ~ 30 m. A lot of tall Clints gather together to form the Stone Forest.

(2) Cone Karst, Hoodoo and Isolated Peak

Cone Karst and Hoodoo are the mountain aggregate formed by limestone suffered from a strong dissolution. Cone Karst is stone peaks which are connected at the bottom of the base; Hoodoo is formed by stone forest which is deep dissolution of the Cone Karst, and only slightly connects to the other on the bottom. Isolated peak is the isolated mountain on Karst area, which is further development of Hoodoo.

(3) Karst Funnel, Uvala and Polje

Karst Funnel is round or oval depression on surface of Karst area. Uvala is the closed depression surrounded by low mountains, hills and Hoodoo. Polje is a large closed depression, which is also known as Dissolution Basin. If the channels on the bottom of the Karst Funnel and Uvala are blocked, water accumulates to come into being the ponds, which may further form large Karst lakes.

(4) Sinkhole, Dry Valley and Blind Valley

Sinkhole is a channel for surface water flowing to the ground or underground

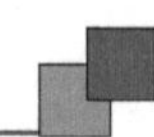

caves on Karst area. It is formed by continuous erosion of vertical Karst to fissure. Sinkhole in the river channel, often makes all the river flows into the underground, which will block the river and form dry valley or blind valley.

2. Underground Configuration

(1) Karst Cave

Karst cave, also called cave, is the underground tunnel resulting from the dissolution and erosion of soluble rock joint or fault by groundwater flow. Karst configurations formed in the Karst caves are stalactites, stalagmites, stone pillars, stone curtain, travertine and sinter. Chinese famous attractions, Anshun Dragon Palace and Zhijin Cave in Guizhou province are the masterpiece of underground Karst landform.

(2) Underground River

Underground river refers to river flowing along horizontal cave under Karst region.

5.3.2 Formation Conditions of Karst

1. Solubility of Rock

Soluble rock is the material basis of Karst development. If the composition of the rock is different, its solubility is different. Carbonate rock has the least solubility, and chlorate rock has the largest solubility. But among the soluble rocks, carbonate rock has the widest distribution, and the same mineral composition. It can all be dissolved by CO_2-containing water. It is the most important stratum for Karst development.

2. Water Permeability of Rock

The water permeability of rock mainly depends on the development degree of gap in the strata, especially the growing degree and the spatial distribution of the fault in rock strata. It can control the growth and distribution of the Karst.

3. Corrosion of Water

Corrosion of water mainly depends on the content of CO_2. The more aggressive CO_2 is in water, the stronger the water dissolution ability will be. A higher CO_2 content will accelerate the dissolution rate of limestone, and hot and humid climate conditions are more favorable for the dissolution.

4. Fluidity of Water

The fluidity of water depends on the circulation conditions of water in the lime rock and groundwater recharge, seepage and drainage. Morphology, scale, density and connection of rock fracture determine the seepage condition of groundwater, and control gradient, flow rate, quantity of flow and flow direction of the groundwater. In addition, topographic slope, nature and thickness of coating layer have an influence on water seepage to some extent.

The main supply of groundwater is atmospheric precipitation. Karst will develop more easily in the area having a higher precipitation.

5.3.3 Prevention and Control of Karst

In order to reduce the engineering workload and ensure the safety of the buildings, the threat of Karst areas should be avoided if the building site can be changed. If Karst areas cannot be avoided, treatments should be taken into consideration. The commonly used engineering treatments are listed as follows:

① Karst water is suitable for dredging, not for blocking. Usually we use open trench and drain cavern to dredge.

② When the cave is buried deeply or has unstable roof, we can use crossing scheme such as long beam foundation, truss foundation or large rigid plate.

③ Digging out the soft fillings in the hole, refilling materials such as gravel, stone or concrete, and compacting hierarchically.

④ When crossing scheme and filling method cannot be adopted, we can grout cement or cement with clay in Karst fissures.

⑤ When cave is buried deeply, we can use pile foundation such as concrete pile, wood pile and sand pile for the reinforcement.

5.4 Earthquake

Earthquake, which is also called the earth motion or earth vibration, refers to a natural phenomenon generated during the process of the rapid release of energy by the Earth's crust. In this process, seismic waves are generated. According to statistics, more than 5 million earthquakes happen on the Earth every year, namely tens of thousands earthquakes happen every day. The majority of earthquake events is too small or too far away, so people do not feel them. About ten or twenty earthquakes can truly cause serious consequences to the human society and one or two earthquakes can cause serious disasters annually. The seismograph records earthquakes that people cannot feel. Different types of seismometers can record earthquakes that have different intensity and different distance. Thousands of seismometers are running around the world, day and night, monitoring the trend of earthquakes.

According to historical records, the highest level of earthquake is a 9.5 degree on Richter scale earthquake in Chile in 1960, which occurred in central sea area of Chile, and triggered a tsunami and volcanic eruptions. The 1976 Tangshan Earthquake in China and the 1995 Hyogoken-Nanbu Earthquake in Japan are the most typical city "straight down type earthquakes" which mean earthquake located in the quakecenter.

5.4.1 Basic Concept of Earthquake

1. Earthquake Epicenter and Focus

(1) Focus

Focus refers to a region where the breaking of interior earth takes place. It is often considered as a point in research work.

(2) Depth of Focus

Depth of focus is the vertical distance from the focus to the ground. The earthquake is called shallow-focus earthquake if the depth of focus is less than 70 km. The earthquake is called intermediate-focus earthquake if the depth of focus is between 70 km and 300 km. The earthquake is called deep-focus earthquake if the depth of focus is more than 300 km.

Most earthquakes are shallow-focus earthquake, with a focal depth between 5 km and 20 km. For example, the focal depth of the 1976 Tangshan Earthquake is 8 km. Intermediate-focus earthquakes is less common, and deep-focus earthquake is rare. The record for the deepest focal depth in the world today is the earthquake happening in east Indonesia Sulawesi Island on June 29, 1934, with a focal depth of 720 km and a magnitude of 6.9.

For the earthquake with the same magnitude, a small area is affected if the focal depth is small, which will cause significant harm. A large area is affected if the focal depth is large, which will cause less harm. Earthquakes with a focal depth over 100 km have nearly no harm on the ground.

(3) Epicenter

The point right on top of the focus on ground is called epicenter, which is considered as an area in practice.

(4) Epicentral Distance

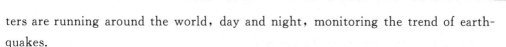

The distance from the epicenter to any point on the ground is called epicentral distance. For the same size of the earthquake, the smaller the epicentral distance is, the greater the damage will be caused.

2. Earthquake Magnitude

Magnitude refers to the energy that an earthquake releases, and it has a fixed value. China's current magnitude standard is the international general Richter scale, which is divided into 10 levels. In actual measurements, the magnitude is calculated according to seismic waves recorded by the seismograph. Earthquakes which are higher than 4.5 magnitudes can be detected around the world.

The released energy differs about 32 times between two consecutive magnitudes. The relationship between the magnitude and the total energy released can be described by the following formula:

$$\lg E = 11.8 + 1.5M$$

where:

E—Energy of earthquake (Unit: erg, 1 erg$=10^{-7}$ J);

M—Earthquake magnitude.

Table 5.2 shows the relation between the earthquake magnitude and energy.

Table 5.3 shows the influence and frequency of the earthquake magnitude.

Table 5.2 Relation between earthquake magnitude and energy

Magnitude	Energy/erg	Equivalent TNT amount	Equivalent instance
0.5	3.55×10^{12}	6 kg	Grenade explosion
1	2×10^{13}	30 kg	Building blaster
2	6.31×10^{14}	1 t	Conventional bomb during the Second World War
3	2×10^{16}	30 t	Large fuel-air bomb in 2003(MOAB)
4	6.31×10^{17}	1 000 t	Small atomic bomb
5	2×10^{19}	33 000 t	Atomic bomb the U.S. dropping at Hiroshima
6	6.31×10^{20}	1×10^{6} t	Double Spring Flat Earthquake in Nevada in 1994
7	2×10^{22}	34×10^{6} t	the most large atomic bomb at present
8	6.31×10^{23}	1.1×10^{6} t	Tangshan Earthquake, Wenchuan Earthquake
9	2×10^{25}	35×10^{9} t	Japan's 9.0 magnitude earthquake in 2011
10	6.31×10^{26}	10^{14} t	The earthquake which is caused by a stony meteorites with a diameter of about 100 km hitting the earth at a speed of 25 km/s

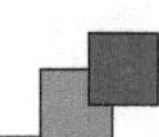

Table 5.3 Influence and frequency of earthquake magnitude

Degree	Richter scale	Earthquake effect	Frequency
Extremely trivial	Under 2.0	Very small, cannot be felt	8 000 times per day
Very trivial	2.0～2.9	People cannot feel, but equipment can detect	1 000 times per day
Small	3.0～3.9	Often have a feeling, but rarely cause damage	49 000 times per year
Weak	4.0～4.9	Indoor shaking occurs, there is unlikely to be a lot of losses, when the earthquake exceed 4.5 on the Richter scale, seismograph can detect it globally	6 200 times per year
Moderate	5.0～5.9	Cause serious damage to deficiently designed and constructed buildings in a small area, but only little damage to well-designed buildings	800 times per year
Strong	6.0～6.9	Destroy residential area within about 100 miles (160 900 m)	120 times per year
Very strong	7.0～7.9	Cause serious damage to the larger area	18 times per year
Extremely strong	8.0～8.9	Destroy area within hundreds of miles	1 time per year
Ultra-strong	Greater than 9.0	Destroy area within thousands of miles	1 time per 20 years

3. Seismic Wave

Focus will shake abruptly when the earthquake occurs. The quake energy will spread out in the form of elastic wave, which is called seismic waves. The seismic wave is called body wave when it is spread in the interior of the Earth. The seismic wave is called surface wave when spreading along the Earth's surface.

(1) Body Wave

Body wave is divided into longitudinal wave and transverse wave.

① Longitudinal wave (P-wave) refers to waves in which the direction of vibration is along the direction of propagation. It can not only spread through solid medium, but also through liquid or gas medium. The propagation speed of longitudinal waves is fast. Longitudinal waves coming from underground will induce the vertical displacement of the Earth's surface.

② Transverse waves (S-wave) refer to waves in which the direction of vibration is vertical to the direction of propagation. It can only spread through solid medium, and the propagation speed is lower than longitudinal waves. Transverse waves coming from underground will induce the horizontal displacement of the Earth's surface. Transverse waves are the main cause of earthquake destruction.

(2) Surface Wave

Surface wave is divided into Love wave and Rayleigh wave. It can only spread through the Earth's surface and has the lowest speed.

① Love wave, also called Q wave, is a kind of surface wave and propagates through multiple internal reflections on the surface layer.

② Rayleigh wave, also called R wave, is a kind of common elastic guided wave. It is a kind of polarized wave that propagates along the free surface of semi-infinite elastic medium.

Because the longitudinal waves propagate faster than the transverse waves in the interior of the Earth, the longitudinal waves always arrive at the ground prior at the transverse waves. Longitudinal waves make the ground deform vertically and transverse waves make the ground deform horizontally. Longitudinal waves arrive at the ground first, shear waves and surface waves arrive later. Transverse waves and surface waves are more intense than the others, which will cause a more serious damage accordingly.

Figure 5.4 shows the propagations of the seismic wave.

4. Seismic Intensity

Seismic intensity denotes how strongly an earthquake affects a specific place such as the ground or infrastructures. After the earthquake, a curve that links all sites which have the same condition of damage is called the isoseismic line. Maps with isoseismic line are called isoseismic diagram. Isoseismic diagram can be concentric circles, concentric elliptical or irregular shape. According to the diagram, local damage as well as the energy transmission of the earthquake at various regions can be observed.

Factors affecting the seismic intensity includes earthquake magnitude, focal depth, the epicenter distance, focal mechanism, geological conditions, underground water level and building performance, etc.

(1) Seismic Intensity Scale

In practice, a unified standard to evaluate the intensity is called seismic

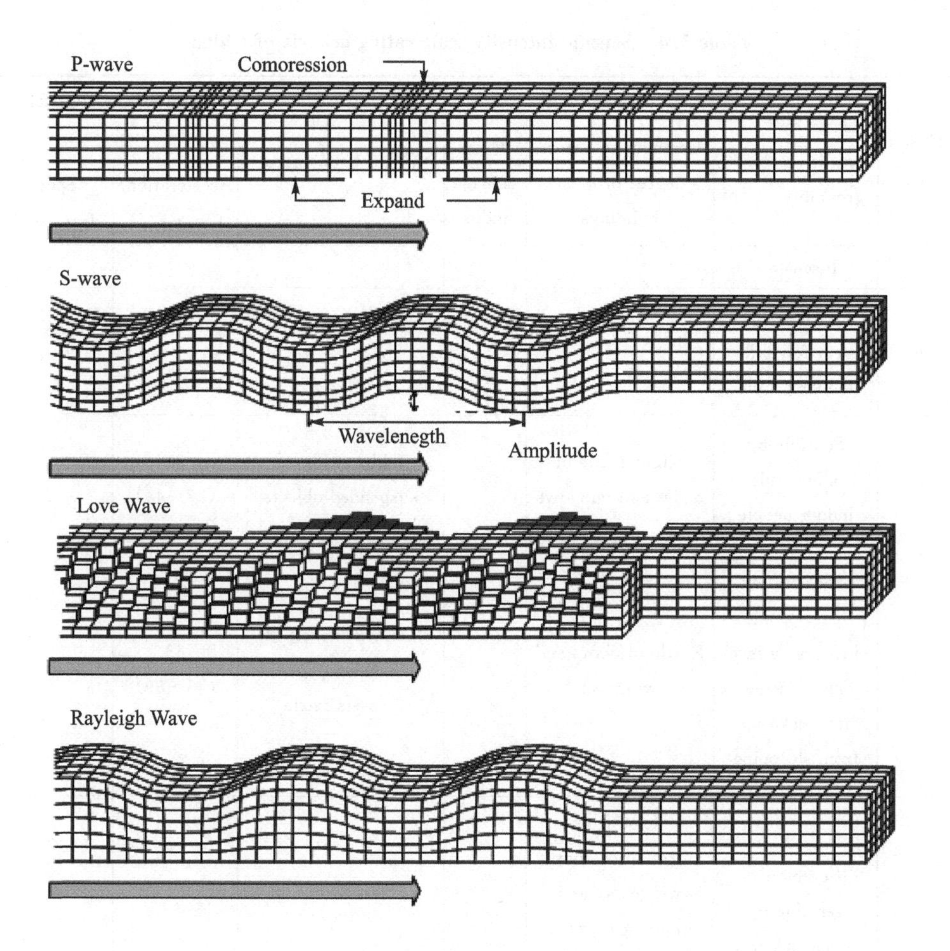

Figure 5.4 Schematic diagram of seismic wave propagation

intensity scale. There are several forms of intensity scale around the world. In 1883, the Italian Rossi and Swiss Frey published RF intensity scale, which divided intensity into IO scales from the trivial tremor to the catastrophe, and used concise description to evaluate the intensity of the macro phenomenon and corresponding symbols, which is widely recognized and adopted. Improved Macquarie intensity scale is very popular in western countries, hereinafter referred to as M. M. Intensity.

Table 5.4 divides intensity into 12 grades from Ⅰ degree to Ⅻ degree. Japan defines non-inductive as zero grade, and divides inductive into 8 grades from Ⅰ to Ⅶ. The former Soviet Union and China divide intensity into 12 grades. Seismic intensity table made by China in 1980 is shown in Table 5.4.

Table 5.5 shows the matching relation of different intensity definition.

Table 5.4 Seismic intensity scale rating criteria of China

Intensity	Senses by people on the ground	Building		Other phenomena	Reference physical indicators	
		Earthquake hazard degree of most buildings	Mean damage index		Acceleration/ (cm · s^{-2})	Speed/ (cm · s^{-1})
Ⅰ	Insensible					
Ⅱ	Sensible by very few still indoor people					
Ⅲ	Sensible by a few still indoor people	Slight rattle of doors and windows		Sight swing of suspended objects	31 (22～44)	
Ⅳ	Sensible by most people indoors, a few people outdoors; a few wake up from sleeping	Rattle of doors and windows		Obvious swing of suspended objects; vessels rattle	63 (45～89)	
Ⅴ	Commonly sensible by people indoors, sensible by most people outdoors; most wake up from sleeping	Noise from vibration of doors, windows, and building frames; falling of dusts, small cracks in plasters, falling of some roof tiles, bricks falling from a few roof-top chimneys		Flipping of unstable objects	125 (90～177)	3 (2～4)
Ⅵ	A few scared to running outdoors	Cracks in the walls, falling of roof tiles, some roof-top chimneys crack or fall apart	0～0.1	Cracks in river banks and soft soil; occasional burst of sand and water from saturated sand layers; cracks on some standalone chimneys	250 (178～353)	6 (5～9)

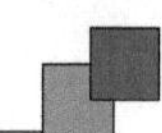

(continued)

Intensity	Senses by people on the ground	Building		Other phenomena	Reference physical indicators	
		Earthquake hazard degree of most buildings	Mean damage index		Acceleration/ (cm • s⁻²)	Speed/ (cm • s⁻¹)
Ⅶ	Most people haste to escape	Slight destruction—localized destruction, crack (continued usage requires small repairs or no repair)	0.11～0.30	Collapse of river banks; frequent burst of sand and water from saturated sand layers; many cracks in soft soils; moderate destruction of most standalone chimneys	500 (354～707)	13 (10～18)
Ⅷ	Most swing about, hard to walk	Moderate destruction—structural destruction occurs, (continued usage requires repair)	0.31～0.50	Cracks appear in hard dry soils; severe destruction of most standalone chimneys	1 000 (708～1 414)	25 (19～35)
Ⅺ	Sit unstably, walking people may fall	Severe destruction—severe structural destruction, localized collapse (difficult to repair)	0.51～0.70	Many cracks in hard dry soils; possible cracks and dislocations in bedrock; frequent landslides and collapses; collapse of many standalone chimneys		50 (36～71)
Ⅹ	Bicycle riders may fall; people in unstable state may fall away; sense of being thrown up	Most collapse and unrepair	0.71～0.90	Cracks in bedrock and earthquake fractures; destruction of bridge arches founded in bedrock; foundation damage or collapse of most standalone chimneys		100 (72～141)

(continued)

Intensity	Senses by people on the ground	Building		Other phenomena	Reference physical indicators	
		Earthquake hazard degree of most buildings	Mean damage index		Acceleration/ $(cm \cdot s^{-2})$	Speed/ $(cm \cdot s^{-1})$
XI		Destruction	0.91～1.0	Earthquake fractures extend a long way; many bedrock cracks and landslides		
XII				Drastic change in landscape, mountains and rivers		

Table 5.5 Matching relation of different intensity definition

Different sorting method	Intensity											
China	I	II	III	IV	V	VI	VII	VIII	IX	X	XI	XII
M. M.	I	II	III	IV	V	VI	VII	VIII	IX	X	XI	XII
Europe	I	II	III	IV	V	VI	VII	VIII	IX	X	XI	XII
RF	I	II	III	IV	V～VI	VII	VIII	IX	X			
Japan	0	I	II	III	IV	V−	V+	VI−	VI+	VII		

(2) Seismic Intensity in Engineering Practice

① Basic intensity, refers to the largest earthquake intensity which a general site may be subjected to in a certain period of time of the future.

② Field intensity, refers to the largest earthquake intensity which a site may be subjected to in the effective use period of the project.

③ Design intensity, refers to the basic intensity that needs to be adjusted according to of safety and financial requirements in engineering design. General buildings can use basic intensity as the design intensity. If there exists adverse site conditions or buildings are important (such as dam, nuclear power station), we can increase the intensity of field appropriately.

5.4.2 Types of Earthquake

1. Classification by Earthquake Magnitude

① Weak earthquake—magnitude<3, generally insensible if earthquake focus is deep.

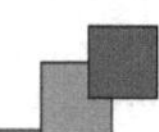

② Sensible earthquake—3≤magnitude≤4.5, sensible but causes no damage.

③ Moderately strong earthquake—4.5<magnitude<6, destructive earthquake. The degree of destruction is related to multiple factors such as the depth of earthquake focus and distance of epicenter.

④ Strong earthquake—earthquake magnitude≥6. When earthquake magnitude is over 8, it is called giant earthquake.

2. Classification by Origin of Earthquake

① Tectonic earthquake—an earthquake caused by huge changes in the tectonic structure because of rupture, displacement and dislocation, which is the main type of earthquake and accounts for 90% in all earthquakes.

② Volcanic earthquake—an earthquake caused by volcanic eruption and crustal shock, which accounts for 7% in all earthquakes.

③ Sinking earthquake—an earthquake caused by ground subsidence, which accounts for 3% in all earthquakes.

④ Inducing earthquake—an earthquake induced by certain external crustal factors in specific areas, such as meteor fall, reservoir filling and water injection in deep wells and so on.

⑤ Artificial earthquake—an earthquake caused by underground nuclear explosion, dynamite blast and other artificial factors.

3. Classification by Degree of Destruction

① Generally destructive earthquake—an earthquake causing several or tens of death or direct economic loss under 1×10^8 RMB.

② Moderately destructive earthquake—an earthquake causing tens or hundreds of death or direct economic loss over one hundred million and under 5×10^8 RMB.

③ Severely destructive earthquake—an earthquake of over 7 magnitude in densely populated area and over 6 magnitude in the big and medium-size cities causing hundreds or thousands of death or direct economic loss over 5×10^8 RMB and under 3×10^9 RMB.

④ Super destructive earthquake—an earthquake of over 7 magnitude or causing tens of thousands death or direct economic loss over 3×10^9 RMB.

5.4.3 Geographic Distribution of Earthquake around the World

It is estimated that there are 85% of earthquakes occurring on the border of continental plates. Earthquakes are contagiously distributed within the earthquake zone and are scattered outside of the zone. The earthquake zones in the world are as follows:

① Pacific rim earthquake zone—distributed around the Pacific, including areas such as the Pacific coast in South and North America, from the north of Aleutian Islands, Kamchatka Peninsula, Japanese Chain Islands and southwards Taiwan Island and through the Philippines southeastwards to New Zealand. It is the most widely distributed and most active earthquake zone in the world, of which the energy released accounts for three quarters of the earthquakes in the world.

② Eurasia earthquake zone—from Mediterranean eastwards. one branch is from middle Asia and the Himalayas and then southwards through Chinese Hengduan Mountains and Myanmar to Indonesia; the other branch is from middle Asia to extend northeastwards to Kamchatka. Earthquakes in this zone are of a scattered distribution.

③ Mid-ocean ridge earthquake zone—meandering through oceans and nearly connecting together, 65 000 km long, 1 000～7 000 km wide and 100 km wide in axial region. This belt has weaker earthquakes than the former two, which are mostly shallow focus earthquake.

④ Continental rifting earthquake zone—smaller in scale than the three above and discretely distributed in inner continents, including East African Rift Valley, Redsea Rift, Baikal Rift and Gulf of Aden Rift and so on.

5.4.4 Influence of Earthquake on Buildings

Earthquake can not only destroy the ground and infrastructures but also cause tsunami, landslide, collapse, fire and flood which are threats to infrastructures.

1. Rupture Effect of Earthquake

When the acting force applied on rock by the seismic waves is larger than the strength of rock, fracture and displacement occur and form dislocation and geofracture in rock, which causes deformation and damages of infrastructures.

2. Effect of Earthquake Liquefaction

Dry and loose silty soil will become compacted by the influence of earthquake shock. When the soil is saturated, the interstitial water pressure in the soil is increasing abruptly. Interstitial water pressure cannot disappear quickly in a short time during earthquake, which makes the effective pressure in grains of sand decrease. When effective pressure disappears totally, the soil loses the shear strength, like liquid, which is called the liquefaction of sand. Liquefaction of sand

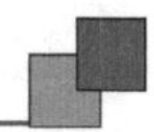

can lead to the falling of ground and distortion by invalid base, even the large scale landslide.

Thinking Questions

1. What are the factors of landslide?
2. Please briefly introduce the measurements of landslide.
3. What are the differences between landslide and collapse?
4. What is debris flow? What are the requirements of debris flow?
5. Please briefly introduce the measurements of debris flow.
6. What is Karst? What are the formation conditions of Karst?
7. What are the forms of Karst?
8. What is earthquake and earthquake magnitude?
9. Please briefly introduce the effects of earthquake on buildings.

Chapter 6

Engineering Geological Investigation

Engineering geological investigation is needed before the city planning and construction to obtain the original information of engineering geology in construction site and to make appropriate working schedule. Besides to achieve the purpose of reasonable exploitation and environmental protection, engineering geological investigation can also avoid the deterioration of geological environment and geological disasters due to the infrastructure construction.

6.1 Tasks and Information Collection during the Investigation Stage

6.1.1 Tasks of Engineering Geological Investigation

Engineering geological investigation is the research work which applies engineering geological theory, geological investigation and testing technology and methods to solve the problems in engineering construction. The result of engineering geological investigation is the important basis for project decision, design and construction. The tasks of engineering geological investigation can be classified as follows:

① To identify the engineering geological conditions of construction sites.

② To identify the age, lithology, geological construction and the distribution pattern of foundation and stratum, and to test physical and mechanical parameters of the rock and soil foundation.

③ To identify the types, qualities, depth and distribution conditions of under-

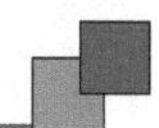

ground water.

④ To analyze the possible engineering geological problems and give reasonable suggestions on the construction forms, foundation type and construction method of building under construction.

⑤ To propose practical resolutions on the rock and soil bed unfavorable for the construction.

⑥ To identify some geological phenomena like landslide and Karst, to analyze the hazardous degree, and to provide the basis for preventing geological disasters.

6.1.2 Contents of Engineering Geological Investigation

① To collect the available information of terrains, landforms, hydrometeor and earthquake and remote sensing photos, engineering experience and investigation reports.

② To investigate engineering geological investigation and survey.

③ To make engineering geological investigation.

④ To test and observe rock and soil.

⑤ To sort out data and write engineering geological investigation report.

6.1.3 Engineering Geological Investigation Stages

The design of construction projects includes three phases: feasibility study, initial design and construction drawing design. The corresponding three stages of investigation work are feasibility study investigation, initial investigation and detailed investigation. As for the foundation of important buildings with complex engineering geological conditions and special construction requirements, a pre-feasibility study and construction investigation are necessary. For construction sites with simple geological conditions and small area, the investigation stage can be simplified.

1. Site Selection Investigation Stage

Site selection investigation is a very important stage for large engineering projects, which is aimed to judge whether the geological conditions of proposed sites are suitable for engineering construction project. In general, several candidate sites will be compared by analyzing the engineering geological information to evaluate the stability and suitability.

Site selection investigation stage should pay attention to the following tasks:

① To collect the geological information of regional geology, such as terrains, landforms, earthquake, minerals, and local construction experience.

② Based on collecting and analyzing the available information, to recognize the underground layer and structure of sites, and properties of rock and soil, and unfavorable geological phenomena and groundwater.

③ To conduct survey and investigation for necessary information for the sites with complex engineering geological conditions and other conditions satisfied.

④ To make budget analysis and generally to avoid the following areas with bad engineering geological conditions:

- Areas with bad geological phenomena in development which pose potential or direct threats to the stability of sites.
- Areas of foundation soil with very bad properties.
- Areas with unfavorable conditions for earthquake resistance.
- Areas with floods and harmful groundwater.
- Areas with underground minerals or unstable underground evacuation sites.

2. Initial Investigation Stage

The purpose of initial investigation is to evaluate the stability of buildings on construction sites to ensure the design plan of general construction layout and foundations as well as to reason the engineering geological conditions for the prevention of first-class bad geological phenomena. The main tasks of this stage are listed in below:

① To collect the feasibility study report of projects and information related to the engineering properties and project scale.

② To initially find out the reasons, distribution range, the influence on stability and developing trend of layers, structures, groundwater and bad geological phenomena and to conduct engineering geological survey and investigation for local complex construction conditions.

③ To evaluate the earthquake effect of construction sites located in areas with earthquake magnitude of 7 or over 7.

In the initial investigation stage, the engineering exploration and test and geophysical exploration are also needed based on collecting information and site conditions.

3. Detailed Investigation Stage

The detailed investigation is made after the completion of initial design. The purpose of detailed investigation is to propose the necessary parameters in design and evaluate the foundation as well as to reason and conclude the specific plan of foundation design and consolidation and the prevention of bad geological phenome-

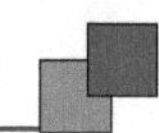

na. The main tasks of detailed investigation stage are as follows:

① To obtain the general construction layout including information of coordinate and terrain, elevation of building, the properties and scale of constructions, possible basic forms and size, anticipated buried depth of the foundation, the unit load and total load of the building, structure properties and special requirements for foundation.

② To find out the reasons, types and distribution, developing trend and hazardous degree of bad geological phenomena, and to propose the technical parameter and plan suggestions of evaluation and rock and soil treatments in need.

③ To find out the types, thickness, engineering properties, stability and bearing capacity of rock and soil in the range of buildings for the calculation and evaluation.

④ For the buildings in need of settlement calculation, it is necessary to give calculating parameters of foundation deformation to predict the settlement, uneven settlement and global tilt of construction.

⑤ For the sites with earthquake intensity of 6 or over 6, it is necessary to classify the types of sites and soil. For the sites with earthquake intensity of 6 and over 6, it is necessary to justify the possibility of liquefaction of saturated sand soil and silty, and to evaluate the degree of liquefaction.

⑥ To find out the deposit conditions of groundwater and to justify the corrosion of groundwater to construction materials. To find out the water level variety range and pattern as well as to provide filtration coefficient of layer for the design of foundation pit.

⑦ To provide the stable calculation of deep foundation pit excavation and geotechnical parameters for the design of supporting structures, and to reason and evaluate the effect of excavation of foundation pit and rainfall on the environment and neighboring projects.

⑧ To provide the geotechnical parameters for the selection of pile types, pile length, pile bearing capacity, pile settlement and construction method.

Detailed investigation is mainly concerned about exploration, in-situ test and lab experiment. Geophysical exploration, engineering geological survey and research can be added if necessary.

6.2 Engineering Geological Mapping

Engineering geological mapping is one of the most important basic survey methods in engineering geology exploration, and it is the first step in exploration work. This method can find out the intrinsic relation between spatial distribution of engineering geological conditions and other elements in the construction area by observing and describing various geological phenomena related to engineering construction using geology and engineering geology theory. Then it faithfully reflected data in the design of a certain scale terrain design drawing in accordance with precision requirements.

Engineering geological mapping is used in feasibility study or the preliminary investigation stage, and we can do some supplementary investigation for some special geological problems in detailed investigation stage.

6.2.1 Main Contents of Engineering Geological Mapping

① Find out the landform characteristics and relationship between these characteristics and the effects of the strata, tectonic structure and geological disaster, and then divide into geomorphic units.

② Find out the formation, origin, age, thickness and distribution of geotechnical properties. As for stratum, we should determine its weathering degree. As for soil layer, we should distinguish between newly sedimentary soils and other kinds of special soil.

③ Research the occurrence, scale and mechanical properties of various structure in testing zone, make sure engineering geological characteristics of all kinds of structures, and analyze its effects on morphology, hydrogeological condition and rock weathering. In addition, we also should pay attention to the characteristics of the new tectonic activity and its relationship with earthquakes.

④ Identify the type, supply sources, drainage and runoff condition of groundwater as well as the location of the wells and springs. Find out aquifer rock properties, buried depth, the water level change and pollution condition as well as its relationship with the surface water, etc.

⑤ Find out the formation, distribution, shape, scale, development degree of some bad geological phenomena such as landslide, debris flow, Karst collapse, gully, fracture, earthquake damage and scouring of shore. In addition, find out the geological disaster's effect on engineering construction. Investigate how the human engineering activities affect the site stability including artificial caves and underground goaf, deep dig and fill, drainage pumping and reservoir induced earthquake etc., monitor building deformation, and collect engineering experience of adjacent constructions.

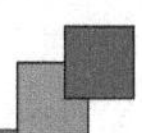

6.2.2 Scope of Engineering Geological Surveying and Mapping

The scope of engineering geological surveying and mapping includes working area and its nearby area. In general, surveying and mapping area should be bigger than building area, but it should not be too big based on the premise of practical problems. Usually we should consider the factors as follows.

1. Building Type

Engineering geology mapping should choose the reasonable scope of mapping depending on the type of building. As for industrial and civil construction, surveying and mapping scope should include building field and its adjacent area. As for road and various lines, surveying and mapping scope should include lines and a certain width of the strip along axis. As for cavity engineering surveying and mapping, the mapping scope should not only include the cavity itself, but also should include the hole in the mountain and its peripheral location. As for reservoir engineering, it should include the area where the geological environment may have a significant change due to the reservoir.

2. Engineering Geological Conditions

If engineering geological conditions are complicated, we should fully find out the engineering geological conditions, and solve engineering geological problems. We can expand the range of surveying and mapping properly, especially we need to consider the dynamic geological process effect. For example, for buildings built near the slope area, the fact that adjacent slopes may produce bad geological phenomena should be taken into consideration when considering scope of surveying and mapping.

6.2.3 Measuring Scale of Engineering Geological Mapping

Measuring scale of engineering geological surveying and mapping mainly depends on the exploration stage, building type, grade, size and the complexity of the engineering geological conditions. Scales that are usually used in engineering geological surveying and mapping are listed as follows.

(1) Reconnaissance and Route of Surveying and Mapping

Scale of 1:200 000 ~ 1:1 000 000, is mainly used to understand the regional engineering geological conditions, and preliminarily estimate building suitability of regional geological conditions.

(2) Small Scale of Surveying and Mapping

Scale of 1:50 000 ~ 1:100 000, is used in geological investigation on feasibility engineering study stage such as highway, railway, water conservancy and hydropower engineering. In industrial and civil construction, underground construction, the scale for this stage is 1:50 000~1:100 000, which is mainly used to find out the engineering geological conditions of the planning area.

(3) Medium Scale of Surveying and Mapping

Scale of 1:10 000 ~ 1:250 000, is used in preliminary design stage of engineering geological investigation such as highway, railway, water conservancy and hydropower engineering. In industrial and civil construction and underground construction, the scale for this stage is 1:20 000 ~ 1:50 000. This is used to find out engineering geological conditions, and get a preliminary analysis of engineering geological stability problems, then provide a geological basis for building the choice.

(4) Large Scale of Surveying and Mapping

Scale that is greater than 1:10 000, is used in drawing design stage of the engineering geological investigation such as highway, railway, water conservancy and hydropower engineering construction. In industrial, civil construction and underground construction we usually use scale of 1:100 to the latter. It is mainly used to find out the engineering geological conditions of building sites, and provide geological basis for selecting architectural form or solving special engineering geological problems.

When engineering geological conditions are complicated, the scale can be appropriately enlarged; we can use enlarged scale in geological elements which has important influence on engineering construction, such as landslide, faults, soft interlayers, caves and so on. On a figure in any scale, line error should not be larger than 0.5 mm.

The detail degree of observation description is controlled by the number of observation points and the length of observation line in each unit area of surveying and mapping. Generally, no matter how large its scale is, we control the average of observing point by making every 1 cm^2 on the map has a breakpoint. When natural outcrop is insufficient, we must add artificial outcrop, so we often need to cooperate with light pitting engineering such as capping, exploratory trench, trial pit in large scale mapping.

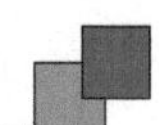

6.2.4 Engineering Geological Mapping Methods

1. Photo Mapping Method

Photo mapping method uses ground satellite photography photos, and in accordance with the indoor judgment sign, combines with the regional geological data, then draws determined formation lithology, geological structure, landform, water system and bad geological phenomena on a single photo, and chooses locations and lines that need to be investigated on that photo. And then, we can do field investigation on the basis of the photo, and check, amend, supplement, finally transfer the results of investigation on the topographic map. Finally, we can get an engineering geological map.

2. Field Surveying and Mapping Method

When the region has no aerial photographs, engineering geological mapping mainly depends on the field surveying and mapping work. There are three kinds of field surveying and mapping methods we commonly use.

(1) Path Method

Along the chosen route and through surveying and mapping field, we draw the surveying and mapping of strata, structure, geology, hydrology, geological boundary and geomorphologic boundary on the topographic map. The route can be a straight line or a broken line. Observation route should be chosen in place where the outcrop and layer is thinner; the direction of observation line is roughly vertical with the strike of the strata, tectonic and geomorphic units. Therefore, we can get more engineering geological data with less effort.

(2) Stationing Method

According to the requirement of the geological condition complexity and scale of surveying and mapping, put a certain number of observation line layouts and observation points on the topographic map in advance. Observation points are generally arranged on the observation line, but considering observation purposes and requirements, such as study geological phenomena geological boundaries, geological structure and hydrological geology, etc. Stationing method is an important method in engineering geological mapping, we often use large and medium scales.

(3) Recourse Method

Stationing points can be set along the boundary of geological structure and geo-

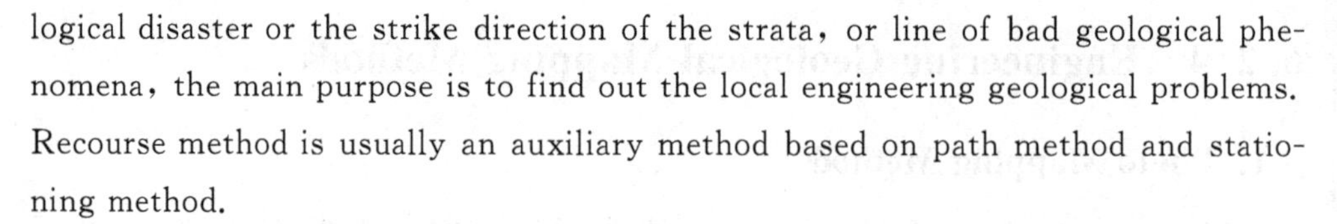

logical disaster or the strike direction of the strata, or line of bad geological phenomena, the main purpose is to find out the local engineering geological problems. Recourse method is usually an auxiliary method based on path method and stationing method.

6.3 The Application of Remote Sensing Technology in Engineering Geological Surveying and Mapping

6.3.1 The Basic Concept

Remote sensing technology: According to the theory of electromagnetic wave, people use various sensing instruments to collect long-range target radiation and reflection of electromagnetic wave information, then process and finally form images. In this way, we can detect and identify various scenes on the ground. It is an integrated measurement technology formed based on the 1960s' aerial photography and interpretation along with the development of space technology and computer technology.

Modern remote sensing technology mainly includes steps such as information acquisition, transmission, storage and processing. Systems that complete all these steps are called the remote sensing system, and the core part of the system is the remote sensor which can obtain information. There are many kinds of remote sensors, such as camera, TV camera, main multispectral scanner, imaging spectrometer, microwave radiometer, synthetic aperture radar and so on. The function of transmission equipment is to transfer sensing information from a far-away platform (for example satellite) back to the station. Information processing equipment includes colour composite apparatus, image interpretation and digital image processor, etc.

According to the platform that carrying remote sensing sensor, remote sensing can be divided into the ground and airborne remote sensing.

① Ground remote sensing: the sensor is installed on the ground platforms, such as car, ship, mobile, fixed or activity elevated platform.

② Airborne remote sensing: the sensor is set on the aircraft, such as balloons, model aircraft and other aircrafts.

③ Space remote sensing: the sensor is set on the spacecraft, such as satellites, spacecraft and space laboratory.

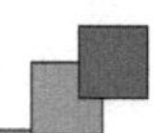

The coverage of remote sensing is quite large, where the working range of aerial photography is about 10 km, and the height of land satellite orbit altitude can reach about 910 km. A Landsat image can cover the above ground area of 30 000 km^2, roughly equivalent to the area of Hainan Island; we only need about 600 Landsat images to cover the whole of China. Remote sensing technology can obtain information quickly and with very short cycle. For example, as for Landsat No. 4 and No. 5, it can cover the Earth once every 16 days, but it may take years even decades to cover the Earth by field of surveying and mapping.

6.3.2 Rationale

Any object has spectral characteristics, which means it has different performances to absorb, reflect and radiate spectrum. In the same spectral region, objects' reflection performances are different, and the reflections of the same object towards different spectra are also different. Even for the same object, in different time and place, due to different sun angle, their reflection and absorption spectra are different. Remote sensing technology can judge objects on the basis of above principles.

Remote sensing technology often uses three regions of the spectrum (green light, red light and infrared light) to detect. Green light is commonly used to detect the characteristics of groundwater, rock and soil; Red light can detect plant growth, change and water pollution, etc. Infrared light can detect land, mineral and resources. In addition, there is microwave section, which is used to detect meteorological cloud and the fish in bottom of the sea.

6.3.3 Application of Remote Sensing Technology in Geological Surveying and Mapping

It will need three stages (preliminary interpretation, field reconnaissance and validation, and mapping) to apply remote sensing data in engineering geological surveying and mapping.

(1) Preliminary Interpretation

According to optical and geometrical characteristics of the geological element on photograph, conduct a systematic observation of aerial photo and stereoscopic, interpret the geomorphology and quaternary geological, divide unconsolidated sediments and bedrock, and do a preliminary structural interpretation work.

(2) Field Reconnaissance and Validation

Factors such as climate, topography, vegetation will make the geological information different. In addition, because of the characteristics of horizon-cover images and remote sensing images, some information is hard to get. Therefore, we need to test in the field in order to supply remote sensing images. At this stage, it is necessary to carry the image into the wild, verify the location of the various typical geological elements in the photograph, select some key research areas and regular spacing through some routes for some actual geologic profile, as well as collect some necessary lithostratigraphic samples.

On-site geological observation points should account for 30% ~ 50% of engineering geological mapping points.

(3) Mapping

Apply the data got from interpretation and field verification as well as other methods to the topographic map, and then do surface structure analysis.

If there is any unreasonable phenomenon, the map should be revised or re-interpretation, when necessary, go to the field and re-inspection, until the whole surface structure is reasonable.

The scale of aerial photo is in the range of 1:250 000 ~ 1:100 000.

Due to the restriction of the bad natural environment and technology level, about 2×10^6 km^2 land of western China doesn't have 1:50 000 scale topographic maps. With the development of surveying and mapping technology, especially the rapid development of photogrammetry and remote sensing technology, the mapping project in western China mainly adopts the space remote sensing, aerial photography, aerospace, synthetic aperture radar, satellite navigation and positioning of photogrammetry and remote sensing technology, such as fast data acquisition technology, high resolution stereo mapping satellite application technology, few control points of remote sensing image topographic survey technology, 24-hour all-weather radar image technology, etc., forming the high-precision topographic map system in western China, which provides strong technical support for the smooth implementation of the mapping projects.

After the Wenchuan Earthquake in 2008, communication and transportation were badly damaged in the disaster areas, and satellite remote sensing and aerial remote sensing technology become the best way to get information quickly. China use optical and radar remote sensing and aerial remote sensing technology to continuously and dynamically monitor the disaster area, carry out interpretation analysis work on disaster area houses, roads and other infrastructure damage, debris flow, landslide, and the lake secondary disasters.

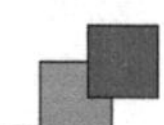

6.4 Engineering Geological Exploration

Engineering geological exploration is a kind of job based on engineering geological mapping and uses certain machine tools or excavation work to find out underground geology work. Exploration methods mainly include drilling, exploratory shaft sinking, trenching and geophysical exploration. Here are several methods commonly used in engineering geology exploration in below.

6.4.1 Drilling

Drilling is a job using drilling machine to drill into the ground to take the core or do geological experiment. It is the most widely used method in exploration. The depth of the engineering geological drilling is usually tens of meters to hundreds of meters, depending on the project requirements and geological conditions, and generally, the depth of civil engineering geological drilling should be less than 10 m. Cylindrical holes that have smaller diameters and a fairly deep depth are called drilling hole. The aperture of drilling holes changes a lot, in general 36 ~ 205 mm; sometimes, we use big aperture drilling hole. Drilling hole whose diameter is larger than 500 mm is called well drilling. The drilling direction is vertical, sometimes inclined, which is called oblique hole. In underground engineering, there are horizontal even vertical upward drilling holes.

1. The Drilling Process

The drilling process has three basic procedures as follows:

① Broken rock and soil, we often use human and mechanical methods, mainly by impact, shear force, grinding and pressure, which make rock or soil separate from whole and become a small number of geotechnical powder.

② Taking rock and soil, we can use rinses or compressed air to rush broken clastic out of hole, or use drilling tools (drawer, spoon drill, twist drill, and soil core tube) remove debris from the bottom of the bore to core surface by human or machine.

③ Preservation of hole wall, in order to drilling go on wheels, we must protect the hole wall against collapse, generally use the casing or slurry supporting.

2. Drilling Methods

Drilling methods can be divided into rotary drilling, impulse type, percussion drilling, and vibration. Each drilling method has different characteristics, which are applicable to different strata.

(1) Rotary Drilling

Rotary drilling is also called the core drilling, refers to drilling under the axis of the pressure breaking the rock by rotary ways. Taking core is optional in the rotary method. Rotary drilling is the main method in drilling rock, in order to keep the core maintain natural state, we usually use clean water as flushing fluid. Rotary drilling can choose different drills made by different materials, such as alloy bit, steel bit and diamond bit. Alloy bit is suitable for drilling soft to medium-hard rock, while steel grit and diamond bit are suitable for drilling hard rock. In order to take rock core of the thin layer, mud, fault or broken rock, we often use double core tube or three core tube to reduce abrasion of rock core. In order to reduce the times of elevating drill pipe and improve efficiency, we also can use rope coring drilling tools. Each time after drilling,the core is taken out of the drill pipe.

Soil holes are not allowed to use commonly rinses rotary drilling, but can use dry drilling.

(2) Percussion Drilling

Percussion Drilling is a drilling method that using drill gravity repeatedly impact on the bottom of the hole and then make soil layer destroyed. Percussion drilling can be divided into human impact and mechanical impact. Human impact drilling tools such as Luoyang hovel, generally are applicable to the shallow hole. Mechanical impact commonly uses lifting and hitting down, is applicable to all kinds of soil drilling. When drilling in the river alluvial gravel layer, in order to obtain samples of sandy gravel, usually adopt the valve pipe impact drilling and the pipe-following drilling.

(3) Impact Rotary Drilling

This is a method combining the impact and rotary drilling. Bore are drilled by the way of rotating under the action of impact of load.

(4) Vibration Drilling

Adopt vibration power produced by the mechanical power, transferring to soils around the cylindrical drill bit by connecting rod and drilling tools, as a result of high speed vibrations of the vibrator. The cylinder bit depends on the weight of the drill and vibrator makes the soil more easily and quickly drilled. It is mainly used in the powder soil, sandy soil, ravel layer having smaller particle size and cohesive soil having small viscosity. The application scopes of various drilling methods are listed in Table 6. 1.

Table 6.1 The application scopes of various drilling methods

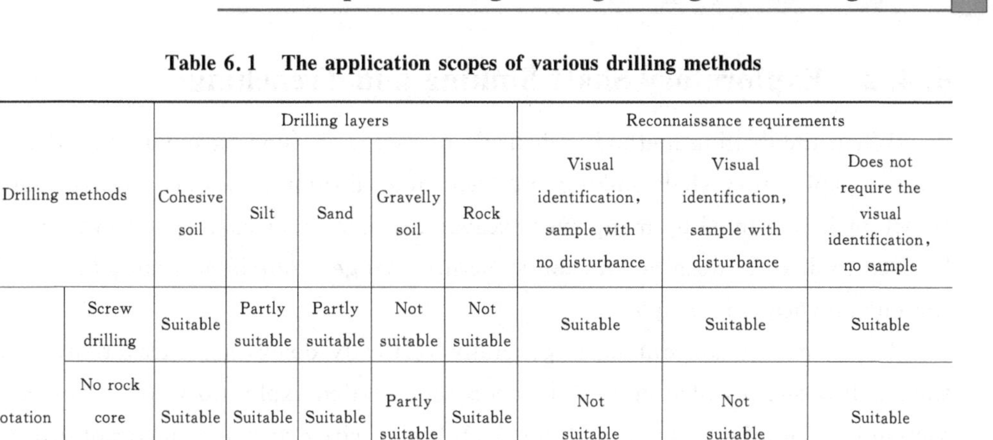

Drilling methods		Drilling layers					Reconnaissance requirements		
		Cohesive soil	Silt	Sand	Gravelly soil	Rock	Visual identification, sample with no disturbance	Visual identification, sample with disturbance	Does not require the visual identification, no sample
Rotation	Screw drilling	Suitable	Partly suitable	Partly suitable	Not suitable	Not suitable	Suitable	Suitable	Suitable
	No rock core drilling	Suitable	Suitable	Suitable	Partly suitable	Suitable	Not suitable	Not suitable	Suitable
	Rock core drilling	Suitable	Suitable	Suitable	Partly suitable	Suitable	Suitable	Suitable	Suitable
Impact	Percussion drilling	Not suitable	Partly suitable	Suitable	Suitable	Partly suitable	Not suitable	Not suitable	Suitable
	Hammer drilling	Partly suitable	Partly suitable	Partly suitable	Partly suitable	Not suitable	Partly suitable	Suitable	Suitable
Vibration drilling		Suitable	Suitable	Suitable	Partly suitable	Not suitable	Partly suitable	Suitable	Suitable

3. Drilling Geological Histogram

Field recording application shall be borne by personnel with professional training, and records shall be true in a timely manner, in accordance with the drilling is presented. The non in-situ records are forbidden.

Drilling field can be identified by naked eye or hand touch method. When there are clear requirements on conditional or survey work, we can use miniature penetrometer quantitative and standardized method, etc. Drilling results can be expressed by drilling geological histogram or layered records. Rock core sample may be retained for a certain time or for a long-term preservation in accordance with the requirements of engineering, or can be made as core and soil core colour photo into the survey results.

Drilling geological histogram is a comprehensive chart including the formation of stratum, diagram includes geological time, soil depth, thickness and the absolute elevation of the bottom of the soil, the description of rock and soil layer, histogram, absolute level ground, underground water level, location of measured date, the selection of rock and soil samples, which has a general scale of 1:100 ~ 1:500.

6.4.2 Exploratory Shaft Sinking and Trenching

When the drilling method is difficult to assess the underground condition, we can use exploratory shaft sinking and trenching. Exploratory well and exploratory trench mainly formed by manpower excavation, also by mechanical excavation, can be used to directly observe stratum structure changes, obtain accurate data or take the undisturbed soil samples.

Exploratory shaft sinking is generally vertically downward excavation, where the shallow one is called pit and the deep one is called exploratory well. The cross-section is generally rectangle of 1.5 m× 1.0 m or circle with a diameter of 0.8 m to 1 m. This method is mainly used to find out the thickness and nature of the coating, sliding surface, cross section, and the underground water level, and take original state soil sample, etc. In the weak soil layer or less cohesive sand and pebble layer, excavation must support the exploratory well and we should pay attention to protect the well mouth, where loose materials cannot be casually discarded in the mouth edge, to avoid the increase of the active earth pressure of borehole wall which could result in borehole wall instability and falling rocks. In the rainy season, we should fortify canopy, dug sewers, to prevent rainwater from infiltrating wall or borehole.

Trenching is a strip drilled on the Earth's surface, usually with a depth of less than 3 m, and a general width of 0.8 ~ 1.0 m. It is used to understand the tectonic line, such as the width of the fault fracture zone, stratigraphic boundary, dike width and extending direction, and take original state soil sample. Trenching shall generally be vertical towards strata or line layout.

In addition to text records, for exploratory shaft sinking and trenching, cross-section diagram including representative colour photos should be used to reflect the bottom rock lithologic, stratigraphic division and tectonic characteristics, sampling and in-situ test position, etc.

6.4.3 Geophysical Prospecting

Geophysical prospecting, uses the principle and method of physics, to observe changes and the distribution of various physical fields, to explore the medium of the ontology and near-earth space structure, material composition, formation and evolution, and to study the change law of various natural phenomena. On this basis, we can detect the Earth's interior structure and tectonics, find theory, method and technology for sources of energy, resources and environment monitoring, and pro-

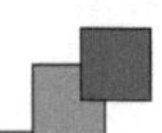

vide important basis for hazard prediction. Geophysical fields include electric field, gravity field, magnetic field, elastic wave field, radiation field, etc.

1. Application Scope of Geophysical Prospecting

Geophysical exploration is an indirect exploration method that can provide results inferring according to the physical phenomena of geologic elements or geological structure. In addition, research on geologic elements or geological structure adopting the geophysical methods is a field source problem based on the measured data or observation of the geophysical field, and is the inversion problem of geophysical field. The result of inversion has lots of solutions, therefore, there are multiple solutions of problems in geophysical prospecting. In order to obtain more accurate and more effective results, we usually use various geophysical exploration methods, and pay attention to combining with geological survey and geological theory to get comprehensive analysis and judgment.

Engineering geological exploration can be used as a means of a leading method of drilling. It is used to understand the hidden geological boundary, interface, or abnormal points. It can also be an auxiliary means of drilling and increase geophysical survey points between boreholes, and provide the basis for the interpolation and extrapolation of drilling results. It can be used as a means of in-situ test, test the rock body wave velocity, dynamic elastic modulus, shear modulus, the characteristics of the cycle, resistivity, radiation parameters and corrosion ability of soil on metal.

When using geophysical prospecting method, we should meet the following two conditions:

① Detected objects have obvious difference of physical properties with the ambient medium.

② Detected objects have certain buried depth and scale, and the geophysical anomaly has enough strength, which can restrain interference, distinguish useful signal and interference signal.

2. Common Methods of Geophysical Exploration

(1) Gravity Prospecting

Gravity exploration is based on Newton's law of gravitation, using the change of the acceleration caused by the density difference of a variety of rock mass of the Earth's crust and orebody surface. As long as there is a certain density difference between exploration geological element and surrounding rock mass, we can use the precise gravity measurement instruments (gravimeter and torsion balance) to find the gravity anomaly. Then, we can do qualitative explanation and quantitative interpretation of the gravity anomaly combining geological and geophysical data in the region, and we can infer layer density under different buried ore body and rock, and then find out the location of existed concealed ore elements and geological structure.

(2) Magnetic Prospecting

Nature of rocks and ores has different magnetism, which can produce different magnetic fields. It results in the change of the Earth's magnetic field of local areas, then generates a geomagnetic anomalies. Magnetic prospecting uses specific instruments to detect and study the magnetic anomaly, and then looks for magnetic ore elements. Magnetic prospecting is one of commonly used geophysical exploration methods, including ground, aviation, marine magnetic prospecting and borehole magnetic measurement, etc. Magnetic prospecting is used to search and explore about mineral (such as iron ore, lead-zinc mine, copper ore and so on), and do geological mapping and research related to oil and gas geological structure and tectonic, etc.

(3) Electrical Prospecting

Electrical prospecting is a method based on the electrical properties of the rock and ore (such as electrical conductivity, electrochemical active properties and dielectric properties, electromagnetic induction, the so-called "electrical differences") to prospect and research on geological structure. It is through the instrument observation of artificial and natural electric field or alternating electromagnetic field, and analyzes, interprets the characteristics of these fields to achieve the purpose of prospecting exploration. Electrical prospecting is divided into two categories. The direct current field is collectively referred to as the direct current method, including resistivity method, charging method, the natural electric field method and direct current induced polarization method, etc. Alternating current (AC) method refers to research of alternating electromagnetic field, including AC induced polarization method, electromagnetic method, the electromagnetic field method, the radio wave perspective method and microwave method, etc. According to the difference of the workplace, electrical prospecting is divided into the ground electrical method, tunnels and well electrical method, aerial electric method, marine electrical method and so on.

(4) Seismic Exploration

Seismic exploration is one of the fastest developed methods in modern geophysical methods, whose principle is to use the propagation law of manmade seismic wave

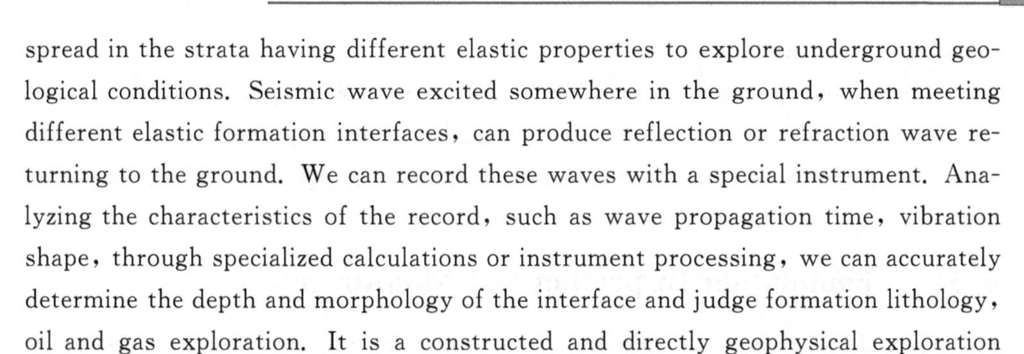

spread in the strata having different elastic properties to explore underground geological conditions. Seismic wave excited somewhere in the ground, when meeting different elastic formation interfaces, can produce reflection or refraction wave returning to the ground. We can record these waves with a special instrument. Analyzing the characteristics of the record, such as wave propagation time, vibration shape, through specialized calculations or instrument processing, we can accurately determine the depth and morphology of the interface and judge formation lithology, oil and gas exploration. It is a constructed and directly geophysical exploration method to find oil, even to find coal field, gypsum deposit, individual hydrogeological and solve engineering geologic layer problems.

6.5 On-site Inspection and Monitoring

Engineering on-site inspection and monitoring is a method that tests the original position of rock and soil layer and maintain its natural structure, natural water content and the natural stress state. It is a supplement with the indoor experiment, and is a complement to each other.

The main methods of in-situ test are load test, static cone penetration test, standard penetration test, vane shear test, compression test and direct shear test. Selection of in-situ test method shall be comprehensively determined according to the requirement of the geotechnical conditions, design requirements, regional experience and the applicability of the test methods and other factors.

After finishing the in-situ test, we need to do field testing and monitoring. On-site testing refers to the check the engineering exploration and inspection results according to the geological conditions exposed in the construction stage. The purpose of the on-site testing is to make the design and construction consistent with the field practice of geotechnical engineering geology, ensure the engineering quality, summarize the experience of exploration, and improve the level of exploration. Field monitoring is a method to monitor the influence of change of rock properties and the surrounding environment conditions in the process of construction. After the completion of construction operation, the purpose of field monitoring is to know the degree of influence caused by the construction and monitor the development trend of this change, in order to take control measures timely during the design and construction. In construction testing and monitoring work phase, if we find that site or the foundation soil conditions have big difference with anticipated conditions, the geotechnical engineering design should be modified or corresponding measures should be taken.

On-site testing and monitoring is an important part in geotechnical engineering. It can not only ensure the engineering quality and safety, improve the efficiency of the project, but also can get some engineering parameters by monitoring means.

6.5.1 Foundation Inspection and Monitoring

Natural foundation pit (groove) inspection is the routine work of geotechnical engineering, and is the last step of the prospecting work. When the construction units finish excavating the base tank, leaders of survey, design, construction units and owner will check groove together on site. Groove check is aimed to examine whether the drilling holes in limited areas are consistent with actual full excavation foundation, whether conclusions and suggestions of survey report are accurate, and solve the problem of new discoveries and problems left in the survey report according to the actual excavation situation of base groove.

Main contents of groove check include whether the excavation plane position of base groove and tank bottom elevation is consistent with exploration and design requirements, whether the load-bearing layer of groove bottom soil is consistent with exploration, and to require staffs to check along the bottom of the channel. When the soil base is significantly uneven or cemetery and grave exist locally, we can use drill rod to find out the range and depth and determine whether the modification or treatments are needed for the foundation scheme.

We usually use simple method such as pocket penetrometer method or naked eye method for the groove check. The portable explorations are also needed when necessary.

① Groove check by observation, we should focus on plinth, corner, main wall subjected to high stress, check the basement soil structure, porosity, humidity and contents etc., and compare with the survey data, to determine whether the designed soil layer has been reached. Local digging inspection should be conducted for suspicious places.

② Groove check by compaction, a method using a wooden tamper, frog-type compaction machine or other construction machines to rammer, pat dry pit (not recommended for wet and soft soil foundation to avoid the destroy of basement soil). Judge whether there is a hole or a grave based on the compaction sound. Further investigation should be taken with portable exploration instruments for suspicious signs.

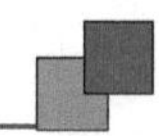

③ Groove check with portable exploration instruments, a method using brazing, lightweight, hand-held dynamic spiral drill, Luoyang shovel to explore on the scope of primary stress layer of foundation, or detect exceptions on the above observation.

6.5.2 Monitoring of Foundation Pit Engineering

The current computational model of the foundation pit engineering and corresponding computing parameters often have discrepancy with actual situation. Therefore, in order to ensure the safety, the monitoring of foundation pit engineering is essential. Through the analysis of monitoring data, if necessary, we can adjust the construction process, and adjust the supporting design. In case of emergency, we should release an alarm in time, and take emergency measures.

From the perspective of the safety of foundation pit, foundation pit monitoring scheme should be determined according to the site conditions and construction design after excavation, and include the deformation of retaining structures, deformation of adjacent construction and underground facilities, underground water level and seepage, flushing and piping, and so on.

Observation of building subsidence can reflect the actual influence on buildings caused by deformation of the foundation, which is an important basis to analyze the foundation accident and judge the construction quality and test the reliability of the survey data, verify the correctness of theoretical calculation of important information.

When we carry out building settlement observation, we should pay attention to the following points:

① Setting of benchmark points have to ensure the stability and reliability, typically set on the bedrock, or set on the soil layer with low compressibility. Location of benchmark points is preferred to be near the observation object, but must be outside the influence range of buildings' pressure. In the same observation zone, the number of benchmark points should not be less than 3.

② Distribution of observation points should fully reflect the deformation of building and is determined together with the geological condition. The number of observation points should more than 6.

③ Precision level and steel rule are usually used for level observation. Same equipment and personnel are preferred for the observation of the same object, and the observation instrument must be strictly checked before leveling. The proper measurement accuracy should be level II, and the length of the sight line is 20 ~

30 m and height of eye line should not be less than 0. 3 m. We should use closure scheme in leveling.

In addition, meteorological data should be recorded at any time in the observation. Observation frequency shall be determined according to the specific circumstances. Under normal circumstances, the civil building should be observed after every construction layer finished. Industrial buildings should be observed according to different load stage, but should not be less than 4 times in the construction phase. Observations of the building after the completion, in the first year, should not be less than 3 ~ 5 times, and should not less than 2 times in the second year, and then once a year, until settlement is stable. For sudden serious cracks or sedimentation, we should increase the frequency of observations.

6.5.3 Monitoring of Adverse Geological Process and Geological Disasters

For the monitoring of adverse geological process and geological disaster, we should compile monitoring scheme based on the geological conditions and engineering practice and then conduct the monitoring according to the defined scheme. The main contents of monitoring scheme include monitoring project, arrangement of measuring points, observation time interval and duration, observation instrument, method and accuracy, data to be submitted, diagrams and disaster forecast and measures to be taken. The following situations should be monitored for adverse geological process and geological disasters:

① When there are bad geological actions or geological disasters near fields, which may endanger the safety of the project or normal use.

② Construction and operation could accelerate the development of the bad geological effect or lead to geological disasters.

③ Construction and operation could have a significant adverse impact on the nearby environment.

6.5.4 Monitoring of Underground Water

Dynamic change of groundwater includes seasonal change of water level and long term change in many years, the change of groundwater caused by human factors, migration of water chemical composition, etc. The monitoring of groundwater is one of the most important key factors in terms of engineering safety and environmental protection. Therefore, the groundwater monitoring has important practical significance. In the following situations, we should carry out the groundwater

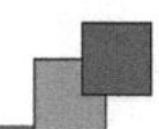

monitoring:

① Rise of underground water level affects geotechnical stability.

② Buoyancy produced by the rise of groundwater level produce big impact on the basement moisture-proof, waterproof, or stability of the underground structures.

③ Construction precipitation has a great influence on proposed project or adjacent construction.

④ Changes of the pore water pressure, groundwater pressure caused by change of construction or environmental conditions, have large influence on engineering design or construction.

⑤ Regional land subsidence caused by increase of the underground water level.

⑥ The rise and fall of groundwater level may make the geotechnical materials soften, collapse and shrink.

⑦ The need for evaluation of environment influence during pollutants migration.

6.6 Interior Work Processing of Survey Data

The final result of engineering geological investigation is the investigation report. When in-situ investigation work and lab experiment are completed, it is necessary to sort out, check, analyze and identify, then to compile the engineering geological investigation report for the design and construction partners. All work above is called the documentation of collected information.

6.6.1 Contents of Engineering Geological Investigation Report

The contents of the engineering geological investigation report should be determined based on mission requirements, the stage of exploration, geological conditions and engineering characteristics, and generally includes the following parts:

① The purpose, task requirement and technical standard accordingly.

② The overview of proposed project.

③ Investigation method and task arrangement.

④ The terrain and landform, layer, geological structure, properties of rock soil and distribution of sites.

⑤ The geotechnical parameter, the suggested value of foundation bearing capacity.

⑥ The deposit condition, types, water level and variation of groundwater.

⑦ The corrosion of water and soil on construction materials.

⑧ Description of potential bad geological effect on stability of projects and evaluation of the harmful degree.

⑨ Evaluation of site stability and suitability.

⑩ Analysis of the utilization and rehabilitation plan of rock and soil materials and proposed suggestions.

⑪ Prediction of possible geotechnical problems during the service period of infrastructures and suggestions on monitoring and prevention.

⑫ Investigation result table and appendix. The types of tables in the appendix should depend on specific engineering conditions. The general tables include exploration site layout, engineering geological columnar profile, engineering geological cross-section profile, in-situ test results table and laboratory test results graphs. If necessary, the appendix can also include comprehensive engineering geological map, generalized geological columnar profile, groundwater level map as well as literary sketch, photos, comprehensive analysis chart, and tables and graphs about geotechnical utilization, treatment and rehabilitation, geotechnical engineering calculation diagrams and calculation result charts.

When necessary, we can also submit the following reports according to the tasks requirement: geotechnical engineering testing report, geotechnical engineering testing or monitoring report, accident investigation and analysis report and report of geotechnical utilization, treatment and rehabilitation methods. Class C geotechnical engineering investigation report can be duly simplified, mainly expressed by diagrams and supplemented by necessary text. Class A geotechnical engineering investigation report can include special testing report, research report or monitoring report on the special geotechnical engineering problems in addition to the fulfill of the above requirements.

6.6.2 Compile of Commonly Used Chart

(1) The Exploration Site Layout

In topographic map, use different representative illustrations to express construction site location, the position of exploration and building test points, and to indicate the exploration, elevation, depth, profile line and serial number of test points.

(2) Drilling Histogram

Drilling histogram is made according to the records of the drilling site. The main contents include the distribution of the foundation soil depth, (thickness of

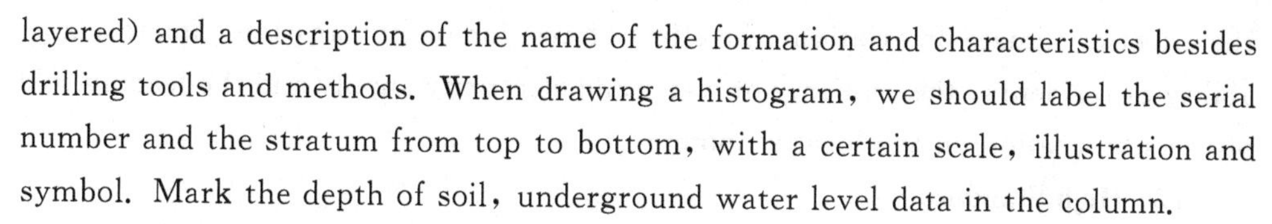

layered) and a description of the name of the formation and characteristics besides drilling tools and methods. When drawing a histogram, we should label the serial number and the stratum from top to bottom, with a certain scale, illustration and symbol. Mark the depth of soil, underground water level data in the column.

(3) Sectional Profile of Engineering Geology

Histogram only reflects the vertical distribution of strata of exploration point, and sectional profile represents a certain exploration line formation distribution along the vertical and horizontal directions. Because prospecting line is vertical to main geomorphic unit or geological structure, or is consistent with the axis of the buildings, the sectional profile can most effectively express site engineering geological conditions.

When drawing the sectional profile, the first step is to draw prospecting line terrain profile line, marking stratum layer in the drilling exploration level, and then to mark the height and depth of layer on both sides of drilling holes. Connect adjacent borehole in the same soil layer by a straight line. When a certain stratum is missed in the adjacent hole, the layer is assumed to be between two adjacent holes. Vertical distance and horizontal distance can be expressed in different scales in the sectional profile.

On the histogram and profile, we can attach the main physical and mechanical properties of soil and some testing curves, such as cone penetration, dynamic penetration, standard penetration test and so on.

6.7 Geological Investigation Requirements in Airport Engineering

6.7.1 Characteristics of Engineering Geological Investigation of Airport Engineering

Morphologically housing infrastructures can be considered as point investigation. Road and railway investigation can be considered as line investigation, and the airport investigation should be the surface investigation. The range of airport detailed investigation is usually over 300 m×2 400 m. Therefore, the airport investigation has the following characteristics.

(1) Strong Systematicness

Besides the suitability of airport engineering conditions, it is also necessary to consider the effect of airport construction on neighboring geological environment,

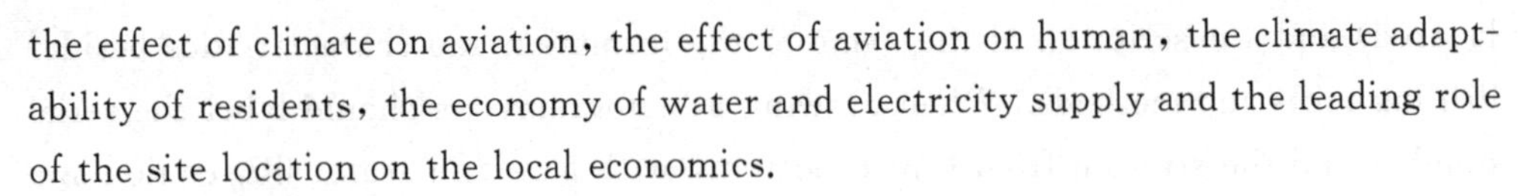

the effect of climate on aviation, the effect of aviation on human, the climate adaptability of residents, the economy of water and electricity supply and the leading role of the site location on the local economics.

(2) High Standard

Airport is the key national investment construction project with requirement of first degree of safety and investigation.

(3) Wide Range of Investigation

The proposed geological investigation area of airport engineering is 10～30 km^2. The area of detailed investigation is 6～15 km^2. The drilling range is over 300 m×2 400 m. There is also a terminal area, and hydraulic and hydropower facilities inspections. The total investigation area is usually over 100 km^2.

(4) Big Challenge for Investigation

Because the aviation zone is required to be a rectangular shape, it cannot avoid river, mud, deep trenches, mountains and other obstacles, often leading to a lack of water and electricity facilities and resulting in the construction difficulties.

6.7.2 Contents of Airport Engineering Geological Investigation

The main methods of airport engineering geological investigation include engineering geological survey, physical exploration, drilling, in-situ measurement and lab experiment. The airport investigation contents include engineering survey and engineering geological investigation. With the increasing requirements of airport construction, the objectives of airport investigation also include hydrology geology, disaster geology, environmental geology and natural building materials.

6.7.3 Classification of Engineering Geological Investigation Phase of Airport Engineering

Airport investigation includes four stages: investigation plan, initial investigation, detailed investigation and construction investigation. The main task of investigation plan is to provide information of several candidate sites for the comparison and determination of selected site. The main task of initial investigation is to conduct initial investigation on the selected site to provide information for the feasibility study of airport. Detailed investigation is to use more methods to conduct investigation with an increased frequency based on the initial investigation, which is aimed to provide information for the construction drawing design. Construction investigation is conducted before or during the construction procedure, which is

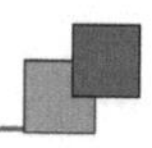

aimed to question detailed investigation or further investigate the problems induced during the design. Construction investigation is usually a small range, local and short-term investigation.

6.8 Geological Investigation Requirements in Civil Engineering

6.8.1 Industrial and Civil Engineering

Residential zone, cultural and sanitary facilities, public facilities and necessary transport lines and pivots, and all kinds of factories and other buildings are mostly shallow foundation structures, where the foundation depth is mostly less than 5 m, the influence depth is just over 10 m and the applied force is mainly static load. Therefore this kind of engineering geological investigation of constructions has the following requirements:

① The investigation objective is the characteristics of terrains, landforms, physical properties of soil and deposit depth and properties of groundwater.

② It is necessary to conduct a large number of shallow holes exploration to analyze the groundwater conditions.

③ Regional planning stage is mainly based on the construction experience and available geological data to reason the overall stability of region.

6.8.2 Road Engineering

Roadwork is the linear project, which usually crosses all of the landforms and geological structures. The stability of road and ordinary service is often influenced by wind, freeze-thaw, subsidence, landslide, debris flow and Karst. Therefore the engineering geological investigation requirements of road engineering are as follows:

① To find out bad geological effect along the route and unfavorable geological conditions for the stability of side slope.

② To obtain longitudinal and cross sectional drawing of the lots under different geological conditions.

③ There is a certain slope limit for the road. On the entire line there are a series of embankment and excavation segments which directly influence the stability of road. Therefore it is necessary to get the relevant parameters based on investigation and experiments, such as types of geotechnical materials, structures, hydrogeological conditions and the material strength.

6.8.3 Underground Engineering

Characteristics of underground engineering are that all structures are imbedded in the underground rock (or soil), so that the safe, economical and proper utilization depends heavily on the stability of surrounding rock. Underground engineering excavation changes the initial stress state of rock and makes surrounding rock loose in a certain range. Without treatments in time, the collapse of underground project will be caused, which could even extend up to the surface and cause the collapse of ground, posing a threat to the safety of buildings.

Common measures to prevent the relaxation of surrounding rocks and falling are tunneling with wooden frame, steel or concrete frames for temporary support during construction and later masoning permanent support structure, known as lining.

In underground engineering excavation, if there exist fracture zone, weathering zone, confined groundwater and other adverse geological conditions, it will cause collapse and water gushing. Therefore in the process of line selection and construction design, a comprehensive understanding of all engineering geological conditions is needed. The requirements of engineering geological investigation are as follows:

① To take various measurements to investigate the lithology, structure of rock media, hydrogeological conditions and geological conditions affecting the stability of mountain, for a purpose of identifying mountain pressure and water gushing.

② In the investigation of preliminary design phase, drilling is indispensable for identifying geological profile. Geophysical methods are encouraged to be adopted if the route lines are too long. If there is a need to determine the physical and mechanical properties of rocks, samples should be taken 20 m above the bottom of the hole. If there is an aquifer above the design elevation of whole bottom, it is necessary to make pumping test to determine the parameters of calculation of water gushing.

③ If engineering geological conditions are not fully identified in the preliminary design stage, supplementary geological survey should be taken. Drilling should be used to further define the nature of rock and geological structure for the tunnel design elevation. In the landslide and fracture zone, or complicated geological conditions such as Karst and thick overburden layer, horizontal investigation lines perpendicular to the vertical axis should be arranged for the preparation of horizontal geological cross-section profile. Investigation guided holes should be arranged (can

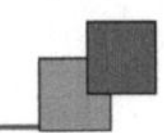

be combined with the construction holes) in the entrance and exit of tunnels, to further identify engineering geological conditions.

Thinking Questions

1. What are the tasks and purposes of engineering geological investigation?

2. What is remote sensing technology? Please name examples to explain the application of remote sensing technique in engineering geology.

3. Please briefly introduce the properties and requirements of airport geological investigation.

References

[1] Bell F G. Engineering Geology[M]. Butterworth-Heinemann,2007.

[2] Shi Zhenming, Kong Xianli. Engineering Geology[M]. Beijing:China Architecture & Building Press, 2013.

[3] Tang Huiming. Principles of Engineering Geology[M]. Beijing:Chemical Industry Press,2015.

[4] Hu Houtian. Civil Engineering Geology [M]. Beijing: Higher Education Press,2001.

[5] Zhang Zhongmiao. Engineering Geolog y [M]. Beijing: China Architecture & Building Press,2007.

[6] Cai Meifeng. Rock Mechanics and Engineering[M]. Beijing:Science Press,2002.

[7] Wang Sijing. Rock Mass Stability Analysis of Underground Engineering[M]. Beijing:Science Press,1984.

[8] Zhang Xiangong,Wang Sijing,Zhang Zhuoyuan. Chinese Engineering:Geology [M]. Beijing:Science Press,2000.

[9] Chen Xizhe. Soil Mechanics and Foundation Engineering[M]. Beijing:China Architecture & Building Press,1998.

[10] Editorial Board of Engineering Geology Manual. Manual of Engineering Geology[M]. Beijing:China Architecture & Building Press,2007.

[11] Drafting Group. Practice Handbook of Geotechnical Engineer[M]. Beijing: China Machine Press,2006.

[12] Code for Investigation of Geotechnical Engineering:GB 50021—2001[S]. Beijing:China Architecture & Building Press,2009.

[13] Survey Specification for City Planning Engineering:CJJ57—2012[S]. Beijing: China Architecture & Building Press,2013.

[14] Geological Survey Specification for Highway Engineering:JTJ C20—2011[S]. Beijing:China Communications Press,2011.

[15] Code for Design of Building Foundation: GB 5007—2011[S]. Beijing: China Architecture & Building Press,2012.

[16] Code for Seismic Design of Building: GB 50011—2010[S]. Beijing: China Architecture & Building Press,2010.